THE LAPWING

THE LAPWING

MICHAEL SHRUBB

Illustrations by
ROBERT GILLMOR

T & A D POYSER
London

First published in 2007 by T&AD Poyser, an imprint of Bloomsbury Publishing Plc.,
50 Bedford Square, London WC1B 3DP

Reprinted 2010, 2012, 2015

www.poyserbooks.com
www.bloomsbury.com

Bloomsbury Publishing: London, New Delhi, New York and Sydney

A CIP catalogue record for this book is available from the British Library

Commissioning Editor: Nigel Redman
Project Editor: Jim Martin

Design by J&L Composition, Filey, North Yorkshire

ISBN (print) 978-0-7136-6854-4
ISBN (epub) 978-1-4081-0898-7
ISBN (ePDF): 978-1-4081-3300-2

Printed and bound in Great Britain

This book is produced using paper that is made from wood grown in managed sustainable forests. It is natural, renewable and recyclable. The logging and manufacturing processes conform to the environmental regulation of the country of origin.

FSC www.fsc.org MIX Paper from responsible sources FSC® C013604

Contents

List of Figures

List of Tables

Introduction and acknowledgements

The Lapwing is perhaps the quintessential farmland bird. As such I have taken a keen interest in its fortunes for many years. Over 90% of the population in Britain breeds on agricultural land and that is also its primary winter habitat. But its status as a farmland bird is, to an important extent, an adaptation. Until the early 19th century it was particularly a bird of what was known as the Waste: semi-natural habitats, used for grazing certainly, but which would hardly be classified as farmland today.

The Lapwing is a beautiful, charismatic and fascinating bird and its loss from so much of the British countryside in my lifetime seems to me, at least, to be something of a tragedy. On our family farm in Sussex, where once it was a common breeding species with 20–30 pairs annually, it now breeds only sporadically. Wintering flocks still occur there regularly but they are only a tithe of what I recorded as a youngster. This was the first bird I ever attempted to do some serious fieldwork on, examining nest success and trying to learn what effect farming practices had upon it in the 1950s. That work taught me how well the Lapwing had learnt to cope with and adapt to classical rotation farming systems. There has never been any doubt in my mind that its loss from so much of the countryside in the past 30–40 years is rooted in its inability to adapt further to the modern revolution in agriculture.

Besides dealing in detail with the Lapwing's breeding and wintering ecology and behaviour, there is inevitably a great deal of agriculture in this book. I have tended to separate patterns of agricultural change in Britain from those in the rest of Europe because it seems clear to me that European experience and history has differed significantly. Denmark appears to me to be the European country where agricultural history and development has most resembled that in Britain. I have also tried to deal with the species over the whole of its range as far as the sources accessible to me allow. But detailed accessible information from the east of the range seems scant outside standard avifaunas. This may be a result of the language barrier as much as anything and may merely reflect my ignorance.

I am grateful to many people for help with this book. First and foremost I must acknowledge the efforts of the many observers whose records have contributed to the national data sets upon which books such as this must draw—wader counts, nest record cards, species surveys, atlas surveys and so on. Then I owe a considerable debt to Carole Showell, the Librarian at the British Trust for Ornithology (BTO), for her unstinting help in obtaining and sending me armies of references. I also have to thank the BTO for access to the maps from the 1998 Lapwing Survey and permission to extract data on dispersion therefrom. The BTO and T & A D Poyser also gave permission to reproduce the map in Figure 13.8 from the Migration Atlas, which Graham Appleton provided. Rob Robinson supplied me with up-to-date data from the EU agricultural statistics, Graham Austin provided the weather data for Figure 5.4, Steve Holloway gave me the national breakdown of WeBS counts for 2001/02 and Rob Fuller kindly supplied me with data from his unpublished winter survey of Lapwings in Buckinghamshire. Rob has also allowed me to use some of his habitat photographs from the same area and Roger Wilmshurst supplied photographs for Plates 8, 11 and 12. Simon Gillings loaned me his Ph.D thesis on the winter ecology of Lapwings in East Anglia, upon which I have drawn extensively. Simon also drew my attention to the work on wintering populations being done in France by Bertrand Trolliet, who very kindly gave me copies of reports and papers deriving from his work. The generous help of all these people is gratefully acknowledged. I am also grateful to Nick Thomas for helpful discussion on agri-environment schemes in Wales, to Mandy Gloyer for similar information for Scotland and to Dr E. Andrews for considerable help on this subject generally.

I would like to thank Robert Gillmor for his splendid drawings, which grace the text so well, and for the paintings on the cover.

Various people have read and commented upon sections of the manuscript. Tony Prater read the chapters on world and European distribution and changes therein, Simon Gillings the chapters on winter distribution and ecology, Mark Bolton the chapters on breeding, Graham Appleton the chapter on migration and Dr E. Andrews that on conservation and the future. Martin Peers read the whole text. I am grateful to Dr Ernest Garcia for the information from the most recent Spanish Atlas; I also thank Nigel Redman and Jim Martin at T & A D Poyser. I am immensely grateful to all these people whose comments, suggestions and corrections have done much to improve the manuscript. Errors and omissions remain my own.

CHAPTER ONE

The lapwing species

The Northern Lapwing was formerly placed in a monotypic genus—*Vanellus*, part of a group of related and distinctive species in the subfamily Vanellinae of the plover family Charadriidae. These species and genera are listed in Table 1.1 but the whole group is now more usually treated as a single genus *Vanellus*, following Sibley & Monroe (1990).

Lapwings are distributed worldwide, except in the Nearctic region. By biogeographical region (Figure 1.1) the greatest concentration of species is in the Afrotropical region, with 11 species breeding: the Spur-winged, Long-toed, Blacksmith, Black-headed, White-headed, Crowned, Senegal Wattled, Senegal, Black-winged, Spot-breasted and Brown-chested Lapwings. The last two have very restricted distributions in the Ethiopian Highlands and West Africa respectively but the remainder are widely distributed south of the Sahara. Three more winter in the Afrotropics: the Sociable, White-tailed and (just) Red-wattled Lapwings. Four species breed only in the Palaearctic region but only the Northern Lapwing is widespread. Sociable, White-tailed and Grey-headed Lapwings occupy restricted

Table 1.1. Historic arrangement of genera and species comprising the subfamily Vanellinae. All species now have the English name 'Lapwing' and are placed in the genus *Vanellus*.

Hoplopterus	Long-toed *V. crassirostris,* Blacksmith *V. armatus,* Spur-winged *V. spinosus,* River *V. duvaucelii,* Black-headed *V. tectus,* Yellow-wattled *V. malabaricus,* White-headed *V. albiceps,* Senegal *V. lugubris,* Black-winged *V. melanopterus,* Crowned *V. coronatus,* Senegal Wattled *V. senegallus,* Spot-breasted *V. melanocephalus,* Grey-headed *V. cinereus,* Red-wattled *V indicus,* Javanese Wattled *V. macropterus,* Banded *V. tricolor,* Masked *V. miles*
Anomalophrys	Brown-chested *V. superciliosus*
Chettusia	Sociable *V. gregarius,* White-tailed *V. leucurus*
Belonopterus	Southern *V. chilensis,* Andean *V. resplendens*
Hoploxypterus	Pied *V. cayanus*
Vanellus	Northern *V. vanellus*

ranges in Central Asia, Central Asia and the Middle East, and in northeast China and Japan respectively. Two other species extend into the Palaearctic, the Spur-winged Lapwing from Africa into the Middle East and southeast Europe and the Red-wattled Lapwing from the Indomalayan region into eastern Arabia, Iraq and Iran. Three species breed in the Indomalayan region, the River Lapwing in Bangladesh and neighbouring parts of India and much of southeast Asia, the Yellow-wattled Lapwing over the bulk of the Indian subcontinent and the Red-wattled Lapwing throughout the Indian subcontintent and southeast Asia. Only the Javanese Wattled Lapwing bred on any Indonesian island but it is almost

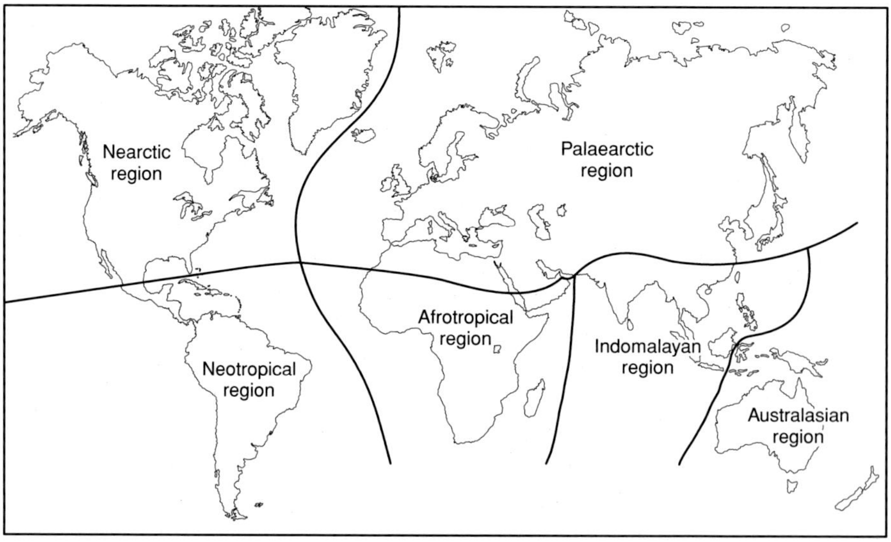

Figure 1.1. The six main zoogeographical regions of the world.

certainly extinct. Northern, Sociable, White-tailed and Grey-headed Lapwings also winter in the Indomalayan region. Two species, the Banded and Masked Lapwings, breed in the Australasian region, the Masked Lapwing being found in New Zealand and New Guinea as well as Australia. Three species breed in the Neotropical region: the Pied, Southern and Andean Lapwings (Vaurie 1965, Cramp & Simmons 1982, Hayman *et al.* 1986, Hagemeijer & Blair 1997).

Although the group is now absent from the Nearctic region except as passage vagrants, the Southern Lapwing has been found in fossil deposits in Florida (Newton 2003).

Lapwings are birds of short grasslands, cultivation and bare ground, often dry, but marshland, pools and river and lake sides are important habitat features and apparently essential to 11 species.

As a group lapwings typically exhibit several common structural and plumage characteristics. Structurally they have broader and more rounded wings than other plovers, a feature which is most marked in the Northern Lapwing. Their flight often therefore lacks the winnowing dash of species such as the golden plovers, although many species have acrobatic display flights and can move very rapidly. It is the character of this looser and more 'floppy' flight that has given rise to the traditional name of lapwing. Although those of the Northern Lapwing are rather short, most lapwings are noticeably long-legged, with the feet extending partly or wholly beyond the tail in flight in 20 species (Table 1.2). Three species have crests, the only waders to do so, 15 have prominent carpal spurs and 11 have facial wattles (eight have both). Only Sociable, White-tailed, Senegal, Crowned and Black-winged Lapwings lack any such adornments. However all lapwings lacking prominent carpal spurs have vestigial ones in the form of bony excrescences under the skin of the carpal joint (Cramp & Simmons 1982).

Lapwings are birds of bold contrast in plumage, with white underwings and most often white underbodies, black primaries and, in many species, bold white bands on the upperwings. Thus even those species which are cryptically coloured at rest are boldly patterned in flight. In tumbling display flights the black/white effect is conspicuous. These bold patterns may also be an effective deterrent against the trampling of nests by large herbivores (Chapter 11). In all species the upper tail is also boldly patterned: white with a bold black terminal band, except in the White-tailed Lapwing, where the tail is all white.

Although the plumage shows sharp contrasts, bright colours are lacking, these being mainly confined to the bare parts. Bills, legs, facial wattles, irides and/or eye rings nearly all exhibit areas of red or yellow (Table 1.2). Only the Sociable Lapwing lacks any such adornment and this species is unique among lapwings in showing a distinct and colourful summer plumage. Seasonal variations in plumage amongst lapwings are otherwise minor and are most marked in Northern, Grey-headed and Yellow-wattled Lapwings.

Movements are varied (Table 1.3). In general those lapwings with breeding ranges extending into cool temperate zones in the northern and southern hemispheres are most strongly migratory, moving away from severe conditions in winter. But several

Table 1.2. Colour of bare parts, leg length and other adornments in adult lapwings.

Species	*Bill colour*	*Leg colour*	*Wattles*	*Iris/eye ring*	*Carpal spur*	*Crest*	*Leg length**
Northern	Black	Dull red	None	Dark	None	Yes	Short
Sociable	Black	Black	None	Dark	None	None	Medium
White-tailed	Black	Yellow	None	Dark/red	None	None	Long
Grey-headed	Yellow/black	Yellow	Yellow	Red/yellow	Yes	None	Medium
Spur-winged	Black	Black	None	Red	Yes	None	Medium
River	Black	Black	None	Red	Yes	None	Long
Long-toed	Red/black	Red	None	Red	Yes	None	Long
Blacksmith	Black	Black	None	Red	Yes	None	Long
Black-headed	Red/black	Red	Pink	Yellow	None	Yes	Medium
White-headed	Yellow/black	Yellow	Yellow	Yellow	Yes	None	Long
Crowned	Red/black	Red	None	Yellow	None	None	Long
Senegal Wattled	Yellow/black	Yellow	Yellow/red	Yellow	Yes	None	Long
Senegal	Black	Dull red	None	Orange/red	None	None	Long
Black-winged	Black	Red	None	Orange/red	None	None	Long
Spot-breasted	Black	Yellow	Yellow	Pale/yellow	Yes	None	Short
Brown-chested	Yellow/black	Brown	Yellow	Yellow	None	None	Medium
Yellow-wattled	Black/Yellow	Yellow	Yellow	Yellow	Yes	None	Long
Red-wattled	Red/black	Yellow	Red	Red	Yes	None	Long
Javanese**	Black	Yellow	Yellow	Yellow	Yes	None	Long
Banded	Yellow	Brown/orange	Red	Yellow	None	None	Short
Masked	Yellow	Orange/red	Yellow	Yellow	Yes	None	Long
Pied	Black	Red	None	Red	Yes	None	Medium
Southern	Red/black	Red	None	Red	Yes	Yes	Medium
Andean	Red/black	Red	None	Red	Yes	None	Short

Data from Hayman *et al.* 1986.

* Under leg length 'short' indicates that feet do not extend beyond the tail in flight, 'medium' that feet protrude and 'long' that the whole foot shows.

** probably extinct.

species are true migrants within Africa, moving in response to the cycle of wet and dry seasons: seasonal shifts in response to rains also occur in some species which are otherwise sedentary. One species, the Masked Lapwing, has no regular migrations but a marked post-breeding dispersal has assisted range expansion.

Five lapwing species breed in the Western Palaearctic: the Northern, Spur-winged, Sociable, White-tailed and Red-wattled Lapwings. Only the Northern Lapwing is generally distributed. The Spur-winged Lapwing breeds in Greece, Turkey, the Levant, the Middle East, Sinai and Egypt and the Sociable Lapwing in Russia and Kazakhstan between 47°N and 53°N, between the Volga and Ural rivers. Apart for occasional records for Turkey, Syria, Armenia, Azerbaijan and the north Caspian coast, the White-tailed Lapwing breeds only in Iraq in this region, although its breeding status there is likely to have been damaged with the drainage of the Euphrates marshes in the 1990s, for this is a species that depends on slow-moving waters. The Red-wattled Lapwing is also confined to Iraq (Cramp & Simmons 1982, Hagemeijer & Blair 1997).

Table 1.3. Patterns of movement in lapwings.

Sedentary	*Seasonal shifts*	*Nomadic*	*Migratory*	*Partial migrant*	*Altitudinal migrant*
Long-toed	Yellow-wattled	Banded	Northern	Black-winged	Andean
Blacksmith	White-headed		Spur-winged*	Senegal Wattled	
River	Crowned		Senegal	Southern	
Black-headed	Red-wattled		Brown-chested		
Spot-breasted			Sociable		
Pied			White-tailed**		
Spur-winged*			Grey-headed		
White-tailed**					

Scientific names are shown in Table 1.1. Sources Cramp & Simmons 1982, Hayman *et al.* 1986, Urban *et al.* 1996.

* African breeding populations of Spur-winged Lapwings are sedentary, European ones migratory.

**Breeding populations of White-tailed Lapwings are sedentary in southern Iraq and Iran but migratory in Central Asia and northern Iraq.

THE NORTHERN LAPWING

The Northern Lapwing, which is hereafter called throughout this book by its older and simpler name of 'Lapwing', is the most numerous and widespread representative of the group in the Palaearctic, breeding right across the region (Chapter 2). It is the only lapwing over much of the Western Palaearctic with a range extending from the Mediterranean countries, where it is more sparsely distributed, to the Arctic Circle and to nearly 70°N in Scandinavia. Over most of this range it is, or was, an abundant species of open countryside, especially agricultural land. Its historical familiarity to country dwellers in Britain is demonstrated by a wealth of vernacular names—Green Plover, Peewit, Puit, Peeseweep, Pyewipe, Bullock-a-week, Teufit, Tewit, Hornpie and Flapjack are a selection: their relationship to call or flight is obvious.

The Lapwing is a beautiful bird. A distinctive medium-sized plover (about the size of a Woodpigeon) of rather stocky build, it looks black and white at any distance. At close range it shows a marked green and, in breeding males, a purple iridescence on the dark upperparts. The underparts are white with a broad black breastband and orange undertail coverts. The wispy crest sweeps up boldly from the rear crown in the male but is shorter and straighter in the female. The wings are rounded and broad, particularly so across the primaries: males have more rounded wings than females. In flight at all seasons it shows plain dark upperparts with white tips to some primaries and a white tail with a broad black terminal band, whilst the underwing is black on the primaries and secondaries but otherwise white. In breeding plumage the adult male's face is black with a black line across white cheeks and a pale grey nape. Females have the black areas on the face and breast variably speckled with white. Seen together, males and females are usually quite easily separable in the breeding season, males being much darker on the back, looking very black and white in many lights (Plates 1 & 2). They are also distinctly larger and, in flight, obviously broader winged across the primaries. My impression is that the wings are longer as well. Females, by contrast, are much duller and browner-green looking on the upperparts and blend in well with the poor unimproved grassland habitats they prefer for nest sites. There is much individual variation in the markings in both sexes which can be useful in studies where individuals need to be recognised.

In winter, adults have a white chin and throat, a buffish nape and face, buffish tips to the mantle and covert feathers and, like breeding females, white specklings to the black breastband. Juveniles are usually distinguishable by their very short crest, buff face, marked pale markings or scalloping to feathers of the back and wings and an incomplete breastband. In flight they show a shorter wing with a distinctly narrower primary area, a difference which is often very obvious in flocks mixed with adults. After their first post-juvenile moult they resemble winter adults. Adults undergo a complete moult from May/June to August/September, occasionally into October, and a partial pre-breeding moult in February/March, occasionally April. The latter involves particularly the feathers of the head, neck and upper breast and change is most marked in males. There is a partial post-juvenile moult in young birds from July-December and a first pre-breeding moult as in adults. Juveniles are indistinguishable from adults after their first post-breeding moult (Cramp & Simmons 1982, Hayman *et al.* 1986, Svensson *et al.* 1999). Despite the extent of the species' range there are no geographical races or, indeed, obvious variations in plumage, probably because of the extent to which populations are mixed in the extensive movements between breeding and wintering grounds (Chapter 13).

Table 1.4 summarises standard measurements and weights. Males are larger than females but there is little variation across the range from which these measurements taken.

The Lapwing's most typical call is often described as plaintive, although it does not sound so to my ear: a fairly high-pitched, disyllabic 'pee-wip', the first syllable more drawn out, the second sharper and more abrupt. It is this call that gives rise to most of the vernacular names, an indication also that it is a very vocal species, with many variations around the basic call, which are described in following chapters where relevant. Its song, delivered in a dramatic tumbling display flight, is a vibrant 'pee-wip wip wip pee-a-wip', to me one of the true sounds of spring. Birds often display and sing on moonlit nights and I know few more evocative sounds than this, ringing across the fields under the moon.

In flight Lapwings show a most distinctive broad-winged silhouette and black-and-white pattern. The flight often looks 'floppy' and rather weak, an appearance which is deceptive in a species which is capable of undertaking regular long-distance migrations, but they are capable of considerable speed when necessary and the broad wings confer a high manoeuvrability, most evident in display.

Table 1.4. Average measurements and weights of Lapwings.

Parameter	*Ad. male*	*Ad. female*	*Juv. male*	*Juv. female*	*Area*	*Source*
Wing (mm)	225	226			Former Soviet Union	Dementiev & Gladkov 1969
Bill (mm)	24.7	23.8			"	"
Tarsus (mm)	49.7	48.3			"	"
Weight (g)	205.3	196.7			"	"
Wing (mm)	229	224	223	221	The Netherlands	Cramp & Simmons 1982
Tail (mm)	106	101	98	97	"	"
Bill (mm)	24.1	23.9			"	"
Tarsus (mm)	47.4	47			"	"
Weights (g)						
Feb.-Mar.	192	189			The Netherlands &Germany	Cramp & Simmons 1982
Apr.-June	211	226			"	"
Sept.-Dec.	254	233			"	"
Aug.-Dec.			195	194	"	"
Mar.-June*			204	203	"	"

*immatures

Lapwings are pre-eminently birds of open country. Today they are usually farmland birds in Europe but that is very largely because farmland is the major use of open countryside. At least in Britain, historical records suggest strongly that farmland was a less important habitat than commons and other features of the Waste (Chapter 4) in the past. They prefer wide open spaces, depending on acute vision to detect danger from predators. Their conspicuously contrasting black-and-white plumage has a important signal function in display and territorial defence (Chapter 9). Graul (1973) also proposed that in the Charadriinae the patterns of bold breastbands and face markings had a disruptive function for nesting birds. Whether the rather similar patterns shown by Lapwings have a similar function I do not know but flocks roosting on broken ground, or birds incubating in similar habitat, can be surprisingly inconspicuous. Outside the breeding season they are social, often occurring in very large flocks in favoured feeding areas. Even as breeding birds they are very often found in loose association or semi-colonies.

CHAPTER TWO

Breeding distribution and populations

Voous (1960) noted that the Lapwing occurred in five climatic zones within the Palaearctic region. It is primarily a bird of the boreal and temperate zones but also breeds in the Mediterranean, steppe and desert climatic zones. Its overall breeding distribution is shown in Figure 2.1. It breeds throughout Europe, including a few pairs in Iceland and the Faroes (Cramp & Simmons 1982, Hagemeijer & Blair 1997). Populations are small in the Mediterranean region (Table 2.1) although it breeds fairly widely in Turkey, mainly across central Anatolia.

In Russia it breeds as far north as Murmansk on the Kola Peninsula, to just north of the Arctic Circle on the Kanin Peninsula, to latitude 65°30′N in the Pechora basin and the range crosses the Arctic Circle again in the southern Yamal Peninsula. The northern boundary of its distribution then tends generally southeastwards to the Sea of Japan, with recent major extensions north of Lake Baikal and in Yakutiya, an isolated breeding area now perhaps joined to the main range, and along the Amur river and its tributaries to 54°N on the Sea of Okhotsk (Tomkovich 1992). A few pairs have recently crossed the Sea of Japan to breed in central Honshu, Japan (Breeding Bird Survey of Japan, 1978).

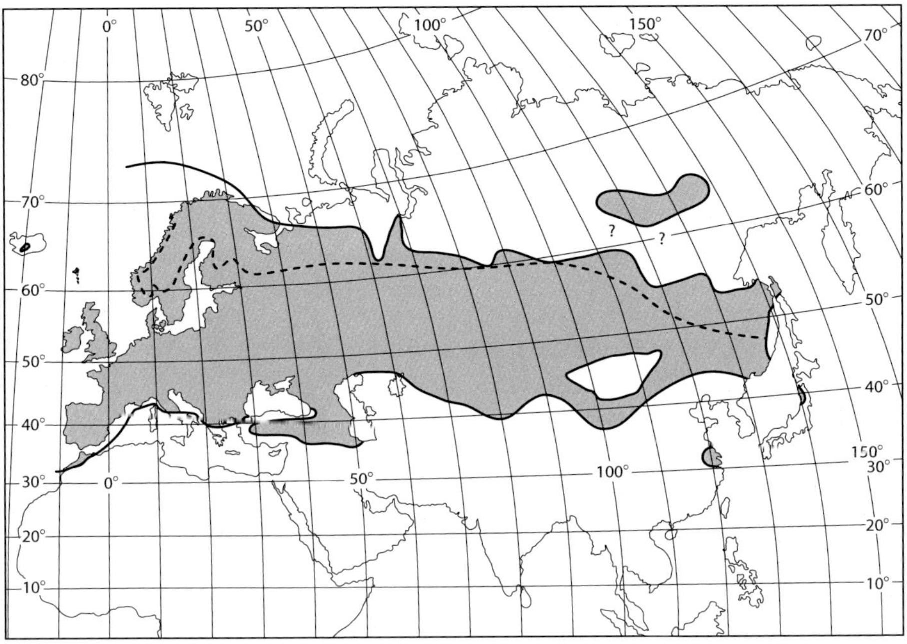

Figure 2.1. The breeding distribution of the Lapwing. For sources see text. The broken line marks the northerly limit of distribution shown by Voous (1960). ? indicates possibly now breeds.

It breeds in Mongolia (Vaurie 1965) and, in China, near the mouth of the Yangtse river but mainly north of 40°N in Manchuria, Inner Mongolia, west to Shensi, Kansu, Tsinghai and Sinkiang (Meyer de Schauensee, 1984). Further west in central Asia it breeds in Kazakhstan south to the mouth of the Emba river on the eastern Caspian and to the north shore of the Aral Sea and south into Kirghizia (Vaurie 1965, Tomkovich 1992). West of the Caspian it breeds south to northern Iran. Generally its southern limits throughout its range lie between about 36°N and 47°N and are largely circumscribed by the presence of extensive desert or mountainous regions.

Populations have been fairly well studied in much of Europe and Table 2.1 gives the most recent known estimate for each country, together with estimates for Armenia, Azerbaijan, Morocco and Turkey. It is important to realize that the quality of these estimates varies and they are likely to be updated as fresh survey information becomes available. In Europe the population is at present considered to total some 1.6 million to 2.8 million pairs, of which 30–40% are found in Russia, two thirds located in central regions. Densities in Russia may not be particularly high overall, the size of the population being at least partly a function of the sheer size of the country.

Figure 2.2 provides another perspective. On the basis of density per unit area of farmland the distribution shows a strong northwesterly bias, well illustrated by

Table 2.1. Breeding populations of the Lapwing in Europe from BirdLife International 2004 and Thorup 2005: the estimate from the most recent survey listed for each country by these authorities is given.

Country	*Number of pairs*	*Country*	*Number of pairs*
Albania	10–45	Macedonia	50–150
Austria	3,000–6,000	Moldova	350–470
Belarus	100,000–160,000	The Netherlands	200,000–300,000
Belgium	17,000–24,000	Norway	40,000–80,000
Bosnia	100–500	Poland	100,000–150,000
Britain	136,000–171,400	Portugal	0–25
Bulgaria	600–1,000	Romania	40,000–60,000
Croatia	3,000–5,000	Russia	500,000–1,100,000*
Czech Republic	7,000–10,000	Serbia	2,000–2,500
Denmark	30,000–45,000	Slovakia	2,800–5,000
Estonia	15,000–30,000	Slovenia	1,500–2,000
Faroes & Iceland	c.12	Spain	1,600
Finland	50,000–80,000	Sweden	50,000–100,000
France	17,000–20,000	Switzerland	250–400
Germany	67,000–104,000	Ukraine	65,000–124,000
Greece	50–100		
Hungary	93,000–150,000	**TOTAL**	**1,592,812–2,788,902**
Ireland	18,770		
Italy	1,700–1,900	Armenia	350–850
Latvia	12,000–15,000	Azerbaijan	500–5,000
Lithuania	18,000–20,000	Morocco	100
Luxembourg	20–30	Turkey	10,000–20,000

Ireland includes Northern Ireland, Austria includes Liechtenstein and Serbia includes Montenegro.

The source for Iceland and Faroes is Hagemeijer & Blair 1997, for Northern Ireland Henderson *et al.* 2002, for Morocco Thévenot *et al.* 2003 and for Spain Martí & Del Moral 2003.

* For Russia, BirdLife International 2004 gives 600,000–1,100,000 pairs and Thorup 2005 gives 500,571–956,405 for the same spread of years. The species is also present in the breeding season in Georgia.

Glutz von Blotzheim *et al.* (1975, Map 56). The very high density in The Netherlands (1,200 pairs or more per 100 km^2 of farmland) may partly reflect a particular feature of the habitat there. This population nests largely in polders, reclaimed from the sea and with an artificially regulated water regime. They have a high water table resulting in an impoverished mammal fauna and a lack of predators (Beintema 1988, Tucker & Evans 1997).

The Lapwing has extended its range and numbers considerably since the mid-19th century, with a substantial expansion along the whole northern boundary of its range. This expansion was particularly marked from the 1940s in Scandinavia and Russia, as the broken line on Figure 2.1, which represents the northern boundary according to Voous (1960), shows. Lapwings have also spread northwest

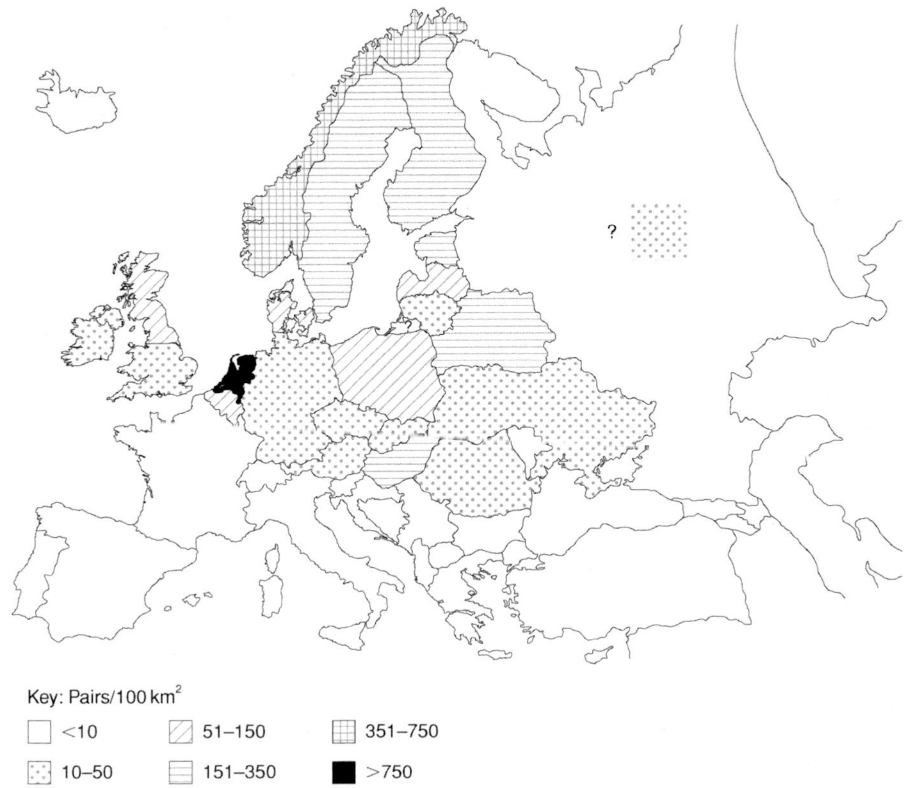

Figure 2.2. Distribution of Lapwings breeding in Europe as pairs/100km² of arable land and pasture. Data from Table 2.1 and Tucker & Evans (1997). Britain is divided into southern England and Wales and northern England and Scotland. Turkey is excluded and species does not breed on the Mediterranean islands.

to the Faroe Islands, where a small population started breeding in the late 1930s (Williamson 1945, Williamson & Boyd 1963), and to Iceland. Here it had become a regular winter visitor by 1951 (Gudmundsson 1951), bred three times after 1963 (Cramp & Simmons 1982) and had three small breeding groups (of 1–9 pairs) during 1985–95 (Hagemeijer & Blair 1997). The small Austrian population was also increasing at this period and Lapwings bred for the first time in Liechtenstein in 1971 (Cramp & Simmons 1982). The main changes are discussed in detail in Appendix 1.

In a few areas, e.g. Iceland and the Faroes, breeding was preceded by the establishment of a wintering population. However, in northern Scotland, Orkney and Shetland, wintering populations did not appear until breeding populations had been long established and perhaps comprised birds from the expanding Scandinavian population (e.g. Venables & Venables 1955).

The colonisation of new breeding areas has often been preceded by a local increase in migrants and breeding range extensions have resulted from migratory flocks overshooting and establishing breeding outposts well north of the existing range. This happened, for example, in the Kola Peninsula in 1939 (Tomkovich 1992) and may well have been the original source of the Shetland population recorded by Saxby (quoted by Venables & Venables 1955). In central Siberia Rogacheva (1992) noted that Lapwings appeared along the Yenisey valley well north of the breeding range virtually every spring, often arriving with warmer spells before the end of April but then departing again with further cold periods. She also recorded that similar increases of overshooting migrants were noted in the Ob river basin, as Lapwings expanded their range there, and along the Vilyuy river in the east, which seem likely to have been the origin of the isolated Yakutiya population.

Recently population increases by the Lapwing have reversed in Western Europe (Appendix 1). Although this decline has yet to alter the species' broad geographic distribution significantly, it is on the verge of extinction outside nature reserves (and not yet safe there) in Wales and southwest England and the scale of recent decline in Northern Ireland suggests a similar situation emerging there. Such declines largely reflect agricultural intensification in Western Europe, and are discussed in greater detail in Chapter 3.

CHAPTER THREE

Breeding habitat and causes of population change in Europe

Lapwing populations have experienced wide-ranging changes in the availability and suitability of their habitats and they have also been affected by more general problems. This chapter discusses how such influences apply across the whole of Europe, including Britain. Additional aspects relating to the Lapwing's fortunes in Britain alone are considered in Chapter 4.

The Lapwing's original breeding habitat was natural grasslands, with short, well-grazed vegetation for nesting and probably often damp areas with shallow pools for fledging chicks. At least in Western Europe such habitats were relatively scarce in prehistoric times, so that Lapwings cannot have been very numerous there. They were probably largely birds of the steppe zone, extending east through southern Russia and central Asia.

As Beintema (1988) remarked, natural grasslands are virtually non-existent in Europe today, having been adapted almost entirely to agricultural use. The expansion of agriculture over four millennia or more also led to the clearance of forest and related habitats and their conversion to farmland on a scale which involved a massive expansion of habitat suitable for the Lapwing, so that it became a common bird of agricultural land throughout temperate Europe. The northern boundary of its range probably always fluctuated with broad changes in climate. It is primarily a lowland species, although it breeds to 1,100 m in southeast Europe (Nankinov 1989). Appendix 2 summarises habitat use in Europe.

With the spread of farming Lapwings came particularly to occupy extensive damp grasslands managed by low intensity grazing systems and for hay crops. Virtually every European account I have examined stresses the importance of wet grassland habitats farmed for grazing and hay crops as core nesting habitats. Outside this habitat type the species' capacity to adapt to agricultural change becomes progressively restricted. Nevertheless it can and does breed in dry habitats such as heaths, dry pastures and arable fields, particularly in areas of mixed spring tillage and grass. The marked recent tendency throughout Europe for the species to spread into arable habitats, shown in Appendix 2, has usually been associated with the loss of wet grassland, although the marked population increase evident until the early 1970s probably contributed. Traditional wet grassland habitats were particularly extensive in Belgium, The Netherlands, north Germany, Denmark and western France. The first four remain core areas for the species in Europe today and grassland management throughout appears to have changed very little until after 1950, except in Denmark (see p.30).

In the Low Countries, particularly The Netherlands, the development of the wet grasslands which hold the most important European population followed a different route to that in other European areas. Israel (1995) noted an important expansion of arable farming there in the 16th century, stimulated by the growth of the industrial cities of Flanders and Brabant. Such expanded agriculture depended on land drainage, but on the peat soils of the region this led to soil shrinkage (as it did later in the English Fens), which was exacerbated further by pumped drainage by windmills in the 17th century. Eventually the fields shrank too close to the groundwater level to allow arable farming and they became wet grassland on which dairy farming developed. With late grazing and mowing imposed by the wet conditions, such farming proved ideal for the emergence of a meadow bird community, including Lapwings (Beintema *et al.* 1997). In addition extensive areas have been reclaimed from the sea since 1945, particularly round the Ijsselmeer and Waddenzee. Such sites were often quickly colonised. Thus van Eerden *et al.* (1979) showed that Lapwings increased on the Lauwerszee area from 20 pairs after reclamation in 1969 to 865 in 1976.

Lapwings defend their nests actively and so need short vegetation with wide fields of view for nesting to detect the approach of predators readily, but they also favour cryptic backgrounds which aid concealment. For rearing, chicks are regularly moved to areas which provide better sources of invertebrate food than many nesting sites (Redfern 1982, Galbraith 1989a). Some management, particularly grazing or mowing, is thus essential in grassland to maintain the habitat in an acceptably open state for nesting, whilst a moderate level of fertilisation enhances food supplies. Abandoning grazing or hay-making leads to grasslands becoming overgrown and eventually colonised by scrub, and useless as Lapwing habitat. Tucker & Evans (1997) noted that recent declines in livestock farming in central and eastern Europe following political transition are reported to have caused widespread changes in the physical structure and species composition of semi-natural grasslands, which are being abandoned. Without grazing such grasslands become rapidly overgrown with coarse grasses and eventually turn into scrub. Such succession can be seen developing, for example, in the extensive hay meadows of eastern Poland (pers. obs.). Important coastal grasslands in the Baltic states are similarly threatened (Tucker & Evans 1997), as happened in the 1960s in the saltmarsh meadows in Scandinavia, formerly extensively grazed by cattle and which were important Lapwing habitats, supporting densities in Finland, for example, of 54–71 pairs/km^2. They are now increasingly deserted by Lapwings with the abandonment of grazing (Soikkeli 1965, Møller 1975, Larsson 1976, Soikkeli & Salo 1979).

There have been two main population trends in Lapwings in Europe in the past 150 years: a long period of increase from the second half of the 19th century until the early 1970s at least, and most marked post-1940, was followed by a sharp decline (Appendix 1). In considering these changes, I have dealt with Britain separately (Chapter 4), as the history of the species and of changes in its habitats there over this period appear to differ significantly from that of continental Europe.

The Lapwing's spread did not occur in isolation. For example, in Finland, von Haartman (1973) noted that 34% of the 233 breeding bird species had increased in range and/or numbers in the previous 100 years, 25% had receded and 20% had stayed stable, the remainder fluctuating. Expanding species included water, marsh and forest birds as well as those of farmland and open country. They also included birds that were expanding west out of Russia and Siberia, as well as those expanding north. Burton (1995), considering the impact of climatic change on birds in the Holarctic, listed 51 species which were retreating northwards during 1900–1950. Clearly bird populations in this period and region were in a markedly dynamic phase, which suggests strongly that some broad environmental factors were operating.

Range expansions on the scale that occurred in Lapwings presumably require successful and increasing source populations, to fill existing habitats and to generate colonists for expansion (Newton 2003). Newton also pointed out that populations at the edge of species' ranges may be limited in different ways to core populations, requiring constant reinforcement to sustain themselves. So range boundaries may fluctuate because of events at the core. Three factors seem likely to have promoted the necessary population growth in Lapwings in Europe from around the 1870s, climatic change, habitat change and protection.

CLIMATE CHANGE

The most significant climatic change was the amelioration of winter temperatures from the second half of the 19th century. Harris (1964) tabulated significant increases in November to March temperatures at eight sites within the Lapwing's range, which rose by an average of 1.82°C between the coldest decade and the warmest between 1870 and 1939. Harris noted that this pattern of rising winter temperatures was common to all European recording stations. March to June temperatures behaved similarly, with an average increase of 1.96°C.

Most probably these climatic changes promoted increasing Lapwing populations through improved winter survival, as Kalela (1949) suggested. A decline in breeding populations after severe winter weather is a well-established phenomenon in Lapwings and, using ringing returns, Peach *et al.* (1994) showed a consistent long-term trend of increasing winter survival in adults during the 20th century, which they noted ultimately represented a 40% increase in the breeding potential of individuals. Adult survival in winter was mainly determined by the severity of the weather, particularly by low temperatures, and was lowest during the most severe winters. The survival of first-year Lapwings in the same period fluctuated markedly without any long-term trend but the lowest survival rates again coincided with severe winters (Chapter 13). Clearly generally milder winters, and perhaps especially the decline in the frequency of severe events, which Williamson (1975) noted had occurred three or four times in each decade in the 19th century, are likely to have been of considerable benefit to breeding Lapwing populations. Unseasonably severe weather in spring can also cause extensive losses in the north. For example a period of frost and snow from 11–17 April 1966 caused declines of 30–90% in breeding Lapwings in south and central Finland (Vepsäläinen 1968: Chapter 13). If rising spring temperatures reduced the frequency of such events, they would have brought similar benefits to rising winter temperatures.

Harris (1964) observed that the trend to milder winters tended to reverse after 1940 and that to milder springs after 1950. Yet the period from about 1940 to the 1970s comprised the period of greatest expansion by Lapwings. Norway provides an excellent example (Figure 3.1). Von Haartman (1973) also showed that the expansion of Lapwings in Finland had significantly outstripped the northerly movement of spring isotherms by 1960, whilst Koskimies (1989) thought that their rapid expansion there was explained by high breeding success and the reserve of yearling birds thus available to colonise or recolonise new areas, ascribing the former to cooperative defence against predators. Nevertheless Peach *et al.* (1994) showed that winter survival continued to improve in this period and these points perhaps underline Williamson's observation that birds may not be particularly precise indicators of climatic change (Williamson 1975).

Climate change may also operate indirectly on Lapwings by its effect on agricultural habitats. Historically at least, it has affected the limits of crop cultivation. Fagan (2000) showed how this shrank in various parts of Europe with

Figure 3.1. Expansion of the range of breeding Lapwings in Norway during the 20th century. Shaded areas represent breeding distribution. Redrawn from Hafthorn (1958) and Gjershaug *et al.* (1994).

the onset of the period known as the Little Ice Age after 1300, whilst Williamson (1975) noted that the limit of crop cultivation moved north by 100 miles in parts of northern Europe and Siberia during the climatic amelioration after 1890 and Varjo (1984) showed how cereal-growing was increasingly limited by northerly latitude in Finland.

Nevertheless I am sceptical that climate change much influenced changes in the main crop areas in northwest Europe during the late 19th and 20th centuries. Williamson (1975) twice particularly commented on the recovery of arable farming in Scotland following climatic amelioration during the first half of the 20th century but this actually misinterpreted the statistics. The areas of both arable farmland (land ploughed in rotation, so tillage plus ley) and tillage (land ploughed annually) declined steadily in Scotland during 1866–1939, falling to their lowest point with economic recession in the 1930s at the peak of the climatic amelioration. They recovered somewhat with the stimulus of wartime demand during 1939–45. It was government and economic policy which influenced these changes, not climate, and the main drivers of agricultural change in northwest Europe since 1800 have been these factors, together with the development of agricultural science and technology. Finland again offers a good example. Varjo (1984) showed that the northerly limits of cereal cultivation moved south by between c.250km (rye) and c.150km (spring wheat) between 1950 and 1975, not because of climatic change but in response to state-managed changes in farming to reduce overproduction.

Climate is not entirely without effect on modern farming, however. A shorter season for cultivations and later harvesting in cooler northern climates, limit the trend to autumn cultivations, an important ecological factor for farmland birds. Climate also limits stocking rates in pastoral farming, since the longer the winter the more land must be set aside to provide winter keep. Agricultural conditions and habitats therefore tend to be more favourable for Lapwings in more northerly regions.

HABITAT CHANGE

Changes in the Lapwing's habitats in Europe in this period generally involved agricultural change. In northwestern Europe changes in agricultural habitats in Scandinavia until 1980 differed sharply from those in the then European Union countries. In Denmark the pattern of agricultural change and trends in Lapwing populations have resembled fairly closely those in Britain, which is dealt with more fully in Chapter 4. With extensive enclosure and drainage of wetlands from the mid-19th century, a mixed farming system similar to high farming in Britain emerged (Møller 1983, Frikke 1991). Throughout the 20th century, however, the trend in Denmark has been to an increase in tilled land, which occupied 77% of agricultural land by 1979, when Denmark joined the Common Market, a far higher proportion than in any other European Union country (Eurostat 1980). Cultivated grassland had declined by 60% during 1900–1984 and the area of permanent freshwater meadow by 82% by 1991 (Larsen 1987, Frikke 1991). As a result of these changes Lapwings declined in Denmark for much of the 20th century and, although they have adapted widely to nesting on arable land, the decline of grassland within the arable area must have undermined the population. Studies at Tøndermarsken showed a strong correlation between the presence of permanent grassland and Lapwing breeding densities (Frikke 1991). Ettrup & Bak (1985) remarked that agricultural areas generally had small populations, which persisted because of immigration from marsh populations, and that the decrease of Lapwings in arable land was probably caused by recent changes in farming practice. Thorup (2005) particularly ascribed it to a massive decline in the area of spring tillage.

In Scandinavia generally, changes in the management of livestock led to the extensive abandonment of traditional shore grazings in pursuit of higher yields from cultivated grassland. The resulting growth of vegetation displaced waders such as Lapwings in favour of passerines (von Haartman 1973, Møller 1975, Larsson 1976, Soikkeli & Salo 1979). In Finland the area of agricultural land also expanded after 1945, with the resettlement of 40,000 farming families from areas ceded to the Soviet Union and other settlers: overall 150,000 new farm units were formed, expanding potential Lapwing habitat considerably (Varjo 1984). Numbers of cattle there also increased by about 37% between 1940 and 1970 but sheep, after a steep increase to 1950, declined overall by 66% in the same period. Stocking rates, however, remained low, tillage was dominated by spring cereals and hay crops occupied 31% of arable land in 1975 (Varjo 1984). Similar changes occurred in Sweden and such agriculture looks very favourable for Lapwings. As the Finnish Lapwing population expanded, smaller fields were colonised (Tiainen *et al.* 1985) and the species spread into areas of peat bog (Hakala 1971). Ratcliffe (2005) noted that peatlands were its main nesting habitat in Lapland, where it bred in small scattered colonies or occasionally as isolated pairs, but he only found it on 6% of the forest peatlands he visited from 1991–2003.

The management of core Lapwing habitats in wet grassland areas within the European Union changed very little until the establishment in 1955 of the Common Agricultural Policy (CAP) of the European Union (then comprising six countries). This stability undoubtedly benefited Lapwings and the overall structure of farmland in fact continued to change rather little between 1955 and 1980, although total agricultural area and arable land both declined. Partly this must reflect the policies and economic arrangements prevailing under the CAP, which promoted uniformity. There was (and remains) a strong tendency for crop areas and so forth to move the same way in all countries involved (Figure 3.2). The main change in agriculture was towards more intensive management, with cereal and grass forage yields rising by 81% and 24% respectively in this period for example.

Two factors, however, particularly affected Lapwing habitats in these countries with the creation of the CAP, the programme of reallocation or consolidation of holdings and the extensive drainage of wet grassland. Reallocation/consolidation

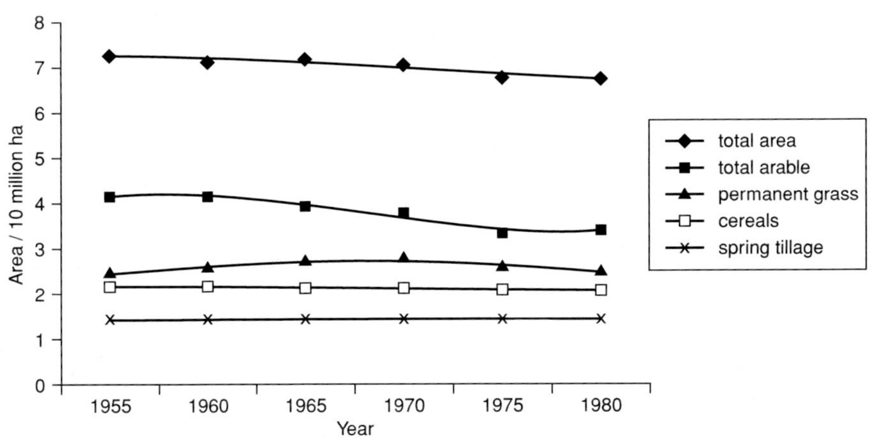

Figure 3.2. Areas of agricultural land in the EU 6 (Germany, France, Italy, The Netherlands, Belgium and Luxembourg) at five-year intervals 1955–1980. Source: Eurostat 1980.

was a process similar in effect to 19th century enclosure in Britain, whereby scattered peasant holdings were reorganised and redistributed into single blocks, usually centred on the owner's dwelling and buildings. Such programmes were instituted to improve the structure of farm units and to obviate the severe economic disadvantages of farming scattered parcels of land at varying distances from the farmstead (Chisholm 1968). Similar programmes were also instituted in Austria, Switzerland, Norway and Finland. They did not directly affect habitat but were accompanied by widespread reorganisation of field systems, with much removal of field boundaries. In France, for example, reallocation affected 72% of farmland after 1950 and involved the removal of two million kilometres of hedgerows (Lefranc 1997). More open landscapes are more attractive to nesting Lapwings, not least because nest survival improves if nests are further from field boundaries and the predators they harbour (Sheldon *et al.* 2004). This process was particularly noted as promoting Lapwing population expansion in Switzerland for example.

Reallocation/consolidation schemes were an important first step for the capital investment needed to intensify European agriculture. In traditional grassland areas in France, Belgium, north Germany and The Netherlands they were followed by extensive drainage and reseeding. It seems likely, however, that the initial impact of these changes on Lapwings was fairly limited and they may have encouraged the species to expand its range into an arable environment which probably remained fairly favourable until the late 1970s.

For Russia and Siberia, Tomkovich (1992) commented that the Lapwing's expansion north along the main river systems of Siberia was related to the development of extensive man-made water meadows in valley bottoms. Presumably these were flood meadows or wet pastures rather than managed water meadows of the type once found on the chalk streams of southern England. He noted also that the species expanded least along the Yenisey, that river having least agricultural improvement. This agricultural development was promoted or organised by the State. Newbury (1980) provided a schematic account of the spread of collectivised agriculture in Siberia from about 1930, where the indigenous people were encouraged to move from a hunting and/or nomadic herding economy to sedentary extensive stock farming. He described a collective farm, the Bala Reindeer Ranch, part of the Verkhoyansk State Farm, c.600km northeast of Yakutsk at about 67°N 134°E. The ranch was about the size of Wales, had lush river valley pastures and carried 800 Reindeer, 700 horses and 600 cows, kept for meat, milk, hides and fur. Some hay was made for winter fodder. Although well north of the Lapwing's present range, this brief description probably gives a good idea of the way the Siberian river systems have been developed agriculturally. A system of management by extensive summer grazing and hay crops in wet grasslands would appear to be ideal habitat conditions for Lapwings and it may be that some of the 20th century expansion in Scandinavia came from the east.

PROTECTION

Newton (2003) discussed the extent to which protection from human persecution and exploitation has led to species increasing and expanding their existing ranges, often, in fact, reoccupying their original ranges. Typical species involved are some of the seabirds, particularly Gannet and Kittiwake in the north Atlantic (Cramp *et al.* 1974).

The Lapwing was a traditional quarry species, exploited both for its eggs and its flesh. Cott (1953), in his survey of the exploitation of wild birds' eggs, lists Britain, The Netherlands and Denmark as the main areas from which Lapwing eggs were gathered more or less systematically for food within the previous 150 years. North Germany was also an important source of eggs imported into Britain (Saunders 1899) and Glutz von Blotzheim *et al.* (1975) noted that egging was an important factor in the species' decline there until 1940. Cott estimated the crop taken annually at 100,000–1,000,000 eggs before the 20th century, when protective measures were increasingly applied. This may have underestimated the scale involved, perhaps considerably, for Newton (1895) noted that 800,000 eggs were imported into London annually from Friesland in The Netherlands alone in the 1870s. Cott lists only seven other species persecuted on a similar scale, Common Eider, Moorhen, Herring Gull, Black-headed Gull, Arctic Tern and Common and Brünnich's Guillemots.

Estimates of the scale of egging for various British sites are summarised in Table 3.1. Being only for particular years and from limited sites such figures only hint at

Table 3.1. Some examples of the scale on which Lapwings' eggs were taken in England and Scotland in the 19th century.

Site	*Quantity*	*Years*	*Reference*
Surrey. Ockley Common	12 doz–clutches/season	Early 20th century	Parr 1972
Kent. Romney Marsh	200 dozen	1839	Yarrell 1845
Suffolk. Thetford	280 doz–annually, one estate	1860s	Ticehurst 1932
Norfolk. Potter Heigham marshes	160 dozen by one egger	1821	Lubbock 1879
Yarmouth, one dealer	600–700 eggs weekly	1830s	Stevenson 1870
Banffshire. One dealer	140 dozen before Apr. 15th	1893	see notes
Tay Valley	40 dozen per week	1896	see notes
Dumfries. Blackshaw	48 dozen in one morning	1888	Gladstone 1910

Notes: Stevenson also noted that the number of eggs reputed to have been taken at East Walton Common, Norfolk, '*is something almost incredible*'. Ticehurst (1932) noted that the crop from the unnamed Thetford estate had declined to 60 dozen annually by the 1880s and to six dozen by 1902. References for Banffshire and Tay Valley are Harvie-Brown & Buckley 1895 and Harvie-Brown 1906.

the scale of the trade. Even so, they indicate that thousands of eggs were gathered for sale in the 19th century and many other similar records are scattered through the literature. For example Baxter & Rintoul (1953) recorded that E.T. Booth took 275 eggs from a ploughed field of less than 1.5ha in East Lothian in 1864.

Such statistics were also gathered mainly at the point of sale and say nothing of the casual collection of eggs for food by farm workers and fishermen and so forth, which was widespread and continued into my boyhood. That such 'casual' collection could be lethal to a local breeding population is illustrated by the history of the Oystercatcher. In both Kent and Norfolk, Ticehurst (1909) and Lubbock (1879) noted that the local breeding populations were exterminated by fishing communities simply taking the eggs to eat themselves. Lubbock noted that the Avocet was similarly affected in north Norfolk.

Nevertheless commercial egging had limitations. It was only profitable where there was a large concentration of breeding Lapwings in a habitat in which they could be easily exploited. For example Tebbutt (1967) and Smith (1887) noted that commercial egging was never viable in Huntingdonshire or Wiltshire. Large concentrations of nesting Lapwings were lacking in the former. In the latter nesting birds were only abundant in arable habitats, where eggs were relatively difficult to find and egging more likely to be discouraged. Such factors underlie the regional limitations on commercial egging that seem evident in the literature. In England it was particularly prevalent in the great marshland and heathland areas of the south and east, which were core areas for the species and, in Scotland, it was again more prevalent in the east than the west, although Glasgow was an important market. In western Britain commercial egging seemed less important, although it certainly occurred in Staffordshire, Cheshire, Ayrshire and Dumfries and probably in Lancashire (Lilford 1895, Gladstone 1910, Paton & Pike 1929, Smith 1930–38). Lilford also noted that eggs were imported into London from Ireland. This minor industry apparently contracted during the 19th century. Yarrell (1845), for example, included Cambridgeshire and Essex in his list of counties providing a large proportion of the supplies to London. However, later authors make no mention of the trade, which suggested that it may have ceased, perhaps because of a decline in the Lapwing population, which Christy (1890) certainly recorded for Essex. In Europe the countries listed above by Cott (1953) were those which, historically, held high densities in wet grasslands. In Friesland, The Netherlands, perhaps the region from which the greatest numbers of eggs were always harvested, egging was sufficiently well-organised in the early 20th century for many villages concerned to appoint guards to prevent poaching after the period when collecting was permitted: at least 100 villages were involved (Brouwer 1952).

Ornithologists have sometimes doubted whether egging, even on this scale, actually affected the population, because of the Lapwing's capacity for laying repeat clutches. Experimental manipulations have induced them to lay up to five clutches in a season (Klomp 1951). Few studies I have examined give any support for the idea that this scale of repeat laying happens outside such experiments and I doubt whether it would obtain in the face of persistent egging over a long period either.

Birds may just give up, as Klinner (1991) found in different circumstances in north Germany (p.41). Matter (1982) also made the point that, in considering the harmlessness or otherwise of egging, attention should be paid to the dynamics of other Lapwing populations. Many breeding on tilled land, particularly in Europe, depend on immigration from more successful populations to persist. It is also important to appreciate the lack of restrictions on collection for much of the 19th century. These did not appear until the 1870s, when there was widespread concern about declining Lapwing populations. Until then egging continued throughout the breeding season and Lubbock (1845) stressed how every egg was taken. Yarrell (1845) described how, in some counties, 'all the most likely ground is carefully searched for eggs, once every day, by women and children, without any reference to the actions of the birds' and quoted Selby that the trade continued for about two months. Knubley (1892) reported that in East Forfarshire (now Angus) 'it is the custom in the nesting season to pay boys 2s per dozen for their eggs, for which 12s 6d is paid to their employers by London dealers, and that no less than fifty dozen are sent off at a time'. Dogs were trained to find nests in Kent and The Netherlands (Ticehurst 1909, Cott 1953) and the literature is full of references to the skill of professional eggers in finding and evaluating nests. Southwell (1879) noted that the development of the rail network in the mid-19th century led to greatly increased demand, particularly from London. Presumably the development of steamships had the same effect on imported supplies. The importance of transport in widening markets was illustrated both by Knubley above and by Stevenson (1870) who noted that 'In the neighbourhood of Holt, some thirty or forty years back, their eggs were taken in considerable quantities . . ., though at that time, from the difficulty of transit, but a small proportion of them reached the inland markets'.

Lapwings' eggs were always valuable and authors from the late 18th century quoted prices of 3–10 shillings per dozen (at least £4.50–£15 in today's money). These were considerable sums at a time when farm workers earned only 10–14 shillings per week. As Lapwings declined and egging was restricted, prices rose and early taken eggs may have commanded as much as 18 shillings per dozen (about £27 today) or more in the 1890s (Cott 1953, see also Knubley 1892 above).

Legislation from the late 19th century has progressively restricted the collection and sale of Lapwings' eggs. In Britain such regulations could be varied by county councils, but only in Ceredigion and the Soke of Peterborough did Lapwing eggs receive no protection by the 1950s (Cott 1953). From the late 19th century legislation limited collection until prescribed dates, usually in the first half of April, but the Lapwing Act of 1926 made it illegal to sell eggs for human consumption or to possess them for the purpose of such sale from 1 March to 31 August. In The Netherlands collection was restricted to before 30 April until 1914, when the close season was moved back to the 28th and in 1937 moved again to the 19th (Brouwer 1952). Collection is still permitted until 12 April in Friesland and until the 5th elsewhere (Beintema *et al.* 1985). One of the reasons that the studies of meadow birds by Beintema and his colleagues was originally commissioned was concern about the effects of continued egg collection on Lapwings. In Germany the population recovery from the 1940s was due to protection and progressive adaptation to nesting on arable land (Glutz von Blotzheim *et al.* 1975).

Egging was not the only exploitation of Lapwings in the past. The bird itself was commonly taken to eat and records from sources such as household accounts in gentry houses, widely quoted in 19th century avifaunas, show it being taken for the table back into the 16th century and earlier. The scale of such exploitation is difficult to judge, at least in Britain. Many of the references to shooting and trapping in the 19th century avifaunas for Britain, for example, are incidental asides, itself an indication that the Lapwing's status as a quarry species was commonplace. Lilford (1895) noted that they 'may easily be decoyed within gun-shot range by means of 'stales' or tethered birds' but remarked that he did not often shoot them, finding them less palatable than Golden Plovers. Nevertheless Kirkman & Hutchinson (1924) included the Lapwing in their survey of British sporting birds and Yarrell (1845) regarded them as excellent for the table. Newton (1895) observed that 'the bird, wary and wild at other times of the year, in the breeding season becomes easily approachable, and is (or used to be) shot down in enormous numbers to be sold in the markets for "Golden Plover". Its growing scarcity as a species was consequently very perceptible—'. Paton & Pike (1929) noted that many were still shot on the Ayrshire coast and Baxter & Rintoul (1953) quoted E.T. Booth that one gunner shot 2,400 Lapwings in three months on the upper Forth in the late 19th century. Cocker & Mabey (2005) noted that large numbers were shot by punt gunners on the fenland washes.

Lapwings are still shot in some numbers in Europe today. Woldhek (1980) gave hunting seasons for a number of Mediterranean countries where shooting was allowed from October until late February or March. Bijlsma *et al.* (1985) noted that hunting pressure on all waders in Portugal was high but Trolliet (2003) noted that Lapwing has not been on the annual List of Huntable Species there for several years, although an open season may still be declared. He noted that Lapwings are now only hunted in four countries in the European Union, France, Spain, Italy and Greece. Elsewhere in the EU it is protected. The largest numbers are shot in France, where Trolliet (2003) recorded that an estimated 1.6 million Lapwings were shot in 1983/84. Since then a decline in the number of hunters and restrictions on the shooting season, which now extends from September to 10 February, have resulted a sharp decline in the numbers shot and Trolliet (2003) noted an estimated 436,000 killed in 1998/99. Trolliet also quoted figures of possibly 100,000 taken in Greece, 200,000–250,000 in Italy and an unknown number in Spain, giving rise to an overall estimate, including birds crippled but not retrieved, of c.1 million birds still being taken. If the present estimate of the winter population given in Chapter 5 is accurate, this represents some 25% of the total, but how accurate the figures for Greece and Italy are is unknown. That for Italy seems too high in relation to the known numbers reported wintering there (Chapter 5). The impact of such shooting may have also declined in recent years, as milder winters are forcing fewer Lapwings into the Mediterranean region, as was also noted by Piersma *et al.* (2005) for Golden Plovers.

In The Netherlands the netting of plovers was a traditional fowling technique, carried on by 'wilstervangers', particularly in the northern provinces of Gronigen, Friesland and Noord-Holland. The main quarry species was Golden Plover but other plovers, and Lapwings in particular, were also taken until the 1940s (Haverschmidt 1943).

The nets used were 20–24m long and 2–3m wide, mounted on poles connected through pulleys by long lines to the operator, concealed behind a blind. Birds were lured into the catching area by stuffed decoys and by live decoys attached to a 'swipe', a long stick which the operator pulled to make the decoys flutter into the air, attracting birds flying over. Call whistles were also used. As birds came in to settle with the decoys the net was pulled over them (Haverschmidt 1943). Haverschmidt pointed out that the method described was the same as that described by Payne-Gallwey (1882) from Ireland. It is also very similar to that described by Stevenson (1870) for the Fens and it seems likely that the method came into eastern England from The Netherlands, perhaps with the Dutch drainers in the 17th century. It clearly has a long history for MacPherson (1897) quoted Thomas Pennant, the 18th century naturalist, describing a similar means of catching Ruff. MacPherson's account also showed that plover-netting was a common pursuit elsewhere in Europe, particularly Italy, where Lapwings were important quarry, and France.

Large numbers of Golden Plovers were taken in The Netherlands. Haverschmidt noted that the 168 netters operating in 1938/39 took 42,000–50,000, which rose to 67,000–81,000 in 1942/43, but he gave no figures for Lapwings. Before 1939 the bulk of the catch was exported, especially to England. Until 1941 large numbers of spring migrants were taken, which was causing a marked population decline. The taking of birds after 15 January was therefore banned from that year.

Stevenson's account shows that Lapwings were much more important quarry in the Fens than Golden Plover. He gave few figures except that one fowler caught 24 dozen in one day: their market price was sixpence (about a pound today) apiece. The method differed from the Dutch in that the fowler set his net and stuffed and live decoys on a small island in a shallow area of flooding, both artificially created. Netting continued into March and April, overlapping with both spring passage and the breeding season. Cocker & Mabey (2005) noted that many netting sites were traditional and recorded that the record morning's catch by one fenman was 240 birds. This was Ernie James, whose obituary in the *Daily Telegraph* (6 August 2005) recorded that he sometimes earned £100 a week selling plovers to dealers before the 1939–45 War. Smith & Cornwallis (1955) noted that thousands were also taken in nets in the Marsh District of Lincolnshire until the use of live decoys was made illegal in 1925. Such trapping continued on a smaller scale, using stuffed decoys, until 1946 when Lapwings were finally fully protected in Lindsey.

As with all wildfowl, a major problem with such exploitation was the lack of any adequate close season. Shooting and netting continued well into the breeding season, extending such activities into the period when it was most damaging to populations (Newton 1998). All the common wildfowl quarry species declined markedly in Britain in the 19th century until a proper close season, starting on

15 February, was imposed in the 1870s. Such problems were by no means confined to Britain.

IMPACT ON POPULATIONS

Beintema (1991), Devos *et al.* (1991) and Tomkovich (1992) all noted that reports of increase and spread involving waders can arise from better counting techniques giving more accurate estimates, or simply from better coverage by more observers. But the expansion of Lapwings in Europe up to the early 1970s was so widely observed and so extensive that its reality cannot be doubted. The problem with attempting to attribute it to any single factor, however, is that the most likely ones discussed here all arose at much the same time. In northwest Europe it seems unlikely that there was sufficient fundamental change in Lapwing habitats in this period to encourage marked range changes, except perhaps in Finland and Sweden. Habitat change tended to be broadly neutral, with losses and gains, but habitat conditions certainly remained broadly favourable to the species. In this region the most likely causes of the Lapwing's marked expansion were a combination of climatic amelioration, encouraging greater winter survival, and declining persecution, with the decline of egging particularly encouraging greater productivity. The timing of these changes matches the timing of expansion well and the progressive nature of expansion is particularly well matched by the progressive tightening of protective legislation. There seems no reason to doubt that the combined impact of these factors would have been sufficient to generate the colonists needed.

In northern Russia and Siberia, by contrast, neither Tomkovich (1992) nor Dementiev & Gladkov (1969) say anything about climatic change or persecution in discussing this species. Nor does Rogacheva (1992) for central Siberia. In these regions the change from hunting and/or nomadic herding economies to extensive stock farming along the valleys of the main river systems from about 1930 created much favourable new habitat, which was probably sufficient to encourage the major northward extension of the species' range there. Tomkovich (1992) particularly stressed the importance of the man-made meadows and remarked that spread into natural habitats, such as saltmarshes and steppe grasslands, was a secondary feature, following colonisation of the water meadows. Both Rogacheva and Tomkovich indicate that Lapwings exhibit a markedly pioneering habit in spring, with overshooting migrants often probing new areas.

RECENT POPULATION CHANGES

The recent history of the Lapwing in northwest Europe has been of a sharp reversal in fortunes. Population estimates listed in Appendix 1 suggest an overall decline of c.35% in the European breeding population during the 1980s and 1990s. The reasons everywhere have been ascribed to agricultural intensification and change. The scale and geographic distribution of this decline is summarised in Figure 3.3. It is clear that it has affected the species throughout its European range. In The Netherlands, which holds the most important west European population, numbers now appear stable but with slow declines in the number using grassland and increases in those on arable land (Appendix 1). The decline has been most marked in Britain (particularly in England and Wales), Denmark, Germany, Finland and Sweden, where it has averaged c.50%. The modern agricultural revolution has many similar features right across northwest Europe, which are dealt with in more detail in Chapter 4 but particular continental European factors are discussed below.

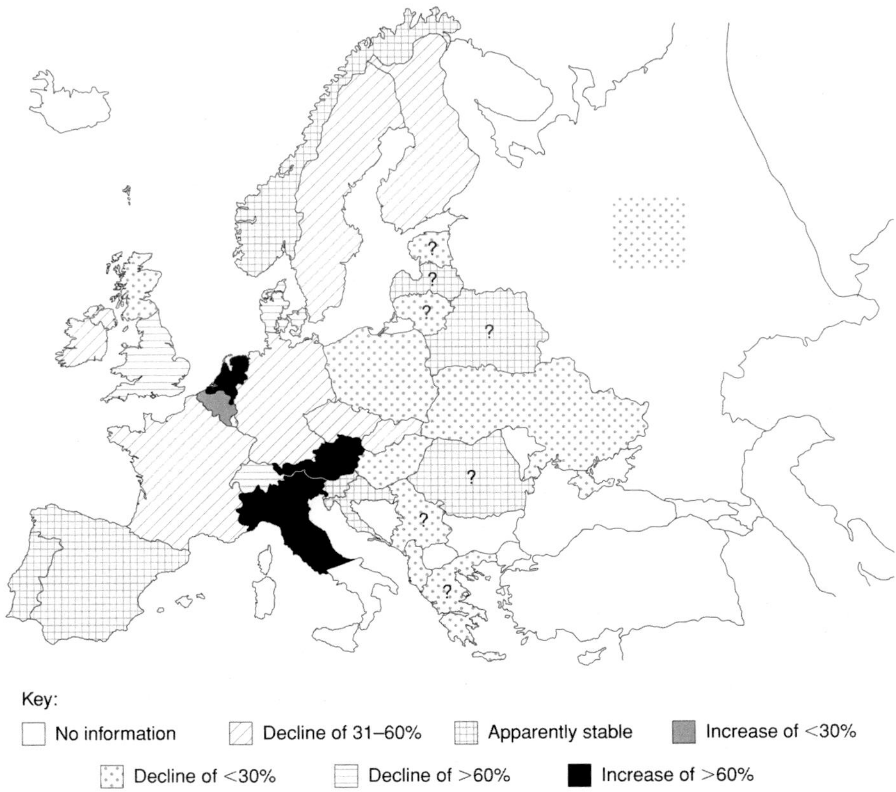

Figure 3.3. Changes in European breeding Lapwing populations during the last 30 years of the 20th century. References and data as in Table 2.1 and Appendix 1. ? indicates change uncertain. Turkey is excluded and species does not breed on the Mediterranean islands.

The agricultural area has declined. Between 1955 and 1995 the agricultural area of the EU 9 (the countries listed in Figure 3.2 plus Denmark, the United Kingdom and Ireland) declined by 11.4 million hectares (Eurostat 1980, 1995) and Meeus *et al.* (1990) noted that about 20 million hectares of farmland were expected to be converted to forest and other uses in the EU 12 (the above plus Spain, Portugal and Greece) by 2000. Similar declines have occurred in Eastern Europe (Tucker & Evans 1997) and in parts of Scandinavia the area of farmland has shrunk even faster. In Finland Varjo (1984) noted that much farmland, particularly in the north, had been abandoned, the number of farms overall declining by c.150,000 between 1969 and 1975. If farm size is fairly uniform (it tends to be larger in the north), that represented nearly 40% of farmland, although there may have been some amalgamation of holdings. Varjo particularly stressed the scale and speed of this contraction in agriculture and also noted that the emphasis on agricultural production has switched to forestry. Both changes are likely to have affected Lapwing habitat, both by direct loss and also, indirectly, by reducing the attractiveness of adjacent farmland, as the species tends to avoid nesting near trees because they harbour predators. Grassland management has also changed. The decline of hay in favour of wrapped silage was very noticeable everywhere in Finland in 2006, and much grassland there looked quite unsuitable for nesting Lapwings, with far too dense a growth. Such changes may well have been sufficient in scale to account for the decline of the Lapwing population in Finland. Trends in Sweden have been similar and Berg (2002) noted that the decline of old grasslands is a major problem for biodiversity there. He also commented that there was a shortage of grazing animals in many parts of Sweden, making the management of grassland habitats difficult. New management regimes, such as grazing every other year, might be introduced to prevent pastures becoming overgrown and to allow some grazing of relatively large areas of pasture.

Whilst decline in the agricultural area involves the loss of potential habitat for Lapwings, a much more significant change has been the widespread drainage of wet grasslands, which represents an extensive loss of core habitat, particularly in Denmark, north Germany, The Netherlands, Belgium and western France. The total area affected runs into many thousands of hectares. As examples, Devos *et al.* (1991) noted the loss of over 30,000ha of wet grassland to drainage for agriculture in Flanders between 1960 and 1987, besides considerable losses to industrial and urban development round Zeebrugge and Antwerp. Busche (1994) noted that 145,000ha of wet grassland had been drained in Schleswig-Holstein between 1950 and 1992 and Frikke (1991) noted that the area of freshwater meadows in Denmark had declined to 46,000ha by 1988, from an area of 250,000ha in 1900 (Larsen 1987): Larsen noted that the rate of loss from 1950–1984 was 2,800ha annually. All round the Baltic, habitat loss to drainage has been compounded by the agricultural abandonment of traditional saltmarsh pastures, once important wader nesting areas, as those that remain still are. In France, Lefranc & Worfolk (1997) noted that 25% of meadows (c.3.5 million hectares) were converted to arable land during 1970–1995.

Perhaps as important as the lowering of water tables in restricting Lapwings has been the agricultural intensification of grassland management which follows. This involves reseeding, earlier grazing, large increases in stocking densities supported by large increases in fertiliser applications, and marked changes in the timing and methods of harvesting forage for winter feed. All these reduce breeding productivity, often to below the level required to sustain populations (Chapter 4).

The predictable result of this intensified grassland management has been to increase production, especially in dairying, to surplus capacity. So production is now restricted by quota, leading to further habitat loss as grassland is converted to tillage. As an example Devos *et al.* (1991) noted that, in Belgium, 3,570ha of grassland per year were converted to arable land between 1977 and 1984 as a result of the restrictions imposed by milk quotas. Lapwings adapt readily to nesting on tilled land, and in Britain arable mixtures of spring tillage and grass are now important core habitats (Chapter 4), but a consistent trend noted in recent breeding studies in Europe has been for populations on arable land to produce too few young to sustain themselves (e.g. Table 5 in Peach *et al.* 1994). Although Beintema (1991) calculated that, provided such populations produced some young they will help to sustain or even increase large populations of waders whilst core areas continue to produce surpluses, the poor breeding success Lapwings now achieve in arable habitats raises the question of why they are so strongly attracted to the habitat for nesting in the first place. Breeding success cannot always have been poor there, otherwise natural selection would have tended to eliminate the habit.

There have also been major changes in the nature of arable habitats throughout much of Europe since about 1970. The most important have certainly been the decline in associated grassland, essential for the successful rearing of chicks, the large increase in winter cereals, particularly wheat which Lapwings avoid for nesting, and the spread of forage or green maize. These crops increasingly dominate tillage in Belgium, The Netherlands, north Germany and Denmark (European agricultural statistics). In The Netherlands, for example, they comprised 85% of tillage by 1990, with green maize occupying 55% (Hustings 1997). In Denmark Thorup (2005) particularly cited the large scale loss of spring tillage as a primary cause of the Lapwing's decline by 80% in farmland.

The spread of green maize has been particularly rapid but its impact has been variable. It results from the increasing trend towards industrial units in European livestock farming, particularly dairying and pig keeping, in which the animals are housed throughout the year. For example, in Denmark, Møller (1983) noted that 25% of cattle were kept permanently in cowhouses by the early 1980s. These large industrial units produce large quantities of waste as slurry, an environmentally hostile by-product. Maize tolerates and can process the quantities of slurry involved but some systems of managing it have proved disastrous for ground-nesting birds such as Lapwings in some areas. Thus, in north Germany, Klinner (1991) noted that maize fields established on what was wet grassland acted as 'ecological traps', more attractive to nesting waders than grass fields, but with cultivations so timed as to catch clutches when incubation was well advanced or chicks had just hatched; slurry spreading was particularly damaging, smothering eggs and chicks. Few chicks survived, increasing numbers of pairs made no attempt to relay after losing clutches and there was also an increasing trend for pairs not to breed at all, simply summering. In consequence, declines of up to 90% in Lapwings were recorded.

Similar problems have been noted in north Wales, where forage maize is an increasingly important crop. Significant numbers of Lapwings are attracted to nest in these fields, which are left bare over winter, but a study in 2000 found that only 17% of nests successfully hatched in maize, compared to 46% in pasture. Females nesting in maize ultimately succeeded in fledging only 0.1 chicks apiece (A.J. Prater pers. comm.). The main problem with growing maize in such areas is the timing of cultivation. As the crop is susceptible to frost damage, cultivation tends to be delayed into late April or early May, so catching nests at a critical stage.

Such problems are not universal, however. At Osnabruck Germany, Kooiker (1984) found that over 80% of nests in maize fields on his study site were successful. Elsewhere maize crops have become an important arable nesting habitat in several European countries, for example in Hungary, Italy and The Netherlands (Appendix 2). As the Hungarian Lapwing population appears to be stable at present and the population in Italy, where maize is a principal nesting habitat, is increasing, it seems rather unlikely that the problems that have emerged in northwest Europe apply there. The difference perhaps reflects differences in spring climate and possibly the fact that grain maize is more important in the south.

Finally, it must be at least possible that the continued high level of losses to hunting in the Lapwing's main winter quarters in France, Spain and Italy is contributing to its decline as a breeding bird in Europe. In an era when breeding performance has declined sharply in the face of habitat loss and agricultural intensification, significant losses to hunting, for which there seems to be very little justification, may be increasingly difficult to sustain.

CHAPTER FOUR

Breeding habitat and causes of population change in Britain

Whilst some of the general points in the previous chapter, particularly the impact of severe weather, egging and persecution, apply equally to Britain and continental Europe, there are marked differences between these regions in the historic patterns of agricultural change and population change in Lapwings. In Britain fundamental habitat changes were brought about by Parliamentary Enclosure, particularly of commons and wastes, from about 1750 to 1870 and with the agricultural recession of the 1920s and 1930s. Furthermore the European population expansion of Lapwings from the late 19th century hardly featured in Britain. The dominant population trend in Lapwings in Britain since the 19th century has been one of decline and what extension of range there was in northwestern Scotland and the Northern Isles happened from around 1840 (Appendix 1) and was perhaps triggered by the rise in soil fertility engendered by the great farming improvements then in progress (Smout 2000, Shrubb 2003).

GENERAL HABITAT USE

In Britain, core Lapwing habitats in the late 18th and early 19th centuries were in the extensive habitats of the Waste—commons, grazing marshes, fens and other wetlands, river floodplains, downland and limestone sheepwalks, brecks, and extensive heaths and moors—all used extensively for grazing and thus as agricultural habitats. Their grazing and other resources were of considerable importance to early rural economies. Such habitats occupied 20% or more of agricultural land in England and Wales and 75% of Scotland in the late 18th century. The uplands have always included large areas of semi-natural grassland but, in this period, every lowland county in England included significant areas of Waste and most parishes had been laid out to provide access to it, a good indication of its value and importance (Tubbs 1993, Shrubb 2003). A greater proportion of the Lapwing population may also have bred on arable land than elsewhere in Europe. Important nesting habitats in southern Britain were the upland cornfields and fallows of the sheep/arable systems on the chalk and limestone (see Shrubb 2003, p.203), for example in Wiltshire (Smith 1887), Oxfordshire (Aplin 1889) and Gloucestershire (Mellersh 1902). Lapwings probably adapted early to this habitat because of the increasing use of the grassland in the river and stream valleys as floated water meadow, a habitat unsuitable for nesting because of the highly intensive management practised. There were 15,000–20,000 acres (6,073–8,097ha) of such meadows in the river and stream valleys of Wiltshire alone (Davis 1794). Overall the literature suggests that the range of habitats occupied by substantial populations was rather wider in Britain than in continental Europe.

There are virtually no counts of breeding numbers in the early literature but Lapwings were certainly very abundant in Britain and Stevenson (1870) wrote that 'At the present day it is only through the 'tales of a grandfather', or the traditionary (*sic*) lore of some octogenarian, that one can arrive at any conception of the former abundance of this species, whose numbers for the last half century, at least, have been gradually but surely decreasing'. The scale of egging shown for the early 19th century in Table 3.2 gives some idea of how numerous this species must once have been.

Table 4.1 summarises broad habitat use by breeding Lapwings in Britain over the past two centuries. It is based on reports in regional and county avifaunas published over the period. Thus 19th century avifaunas provided 89 reports of habitat usage, 65% being for the semi-natural habitats of the Waste (see above), 33% for improved farmland and 2% for habitats outside farmland: such as clay and gravel pits, reservoirs, industrial waste land, lime beds, saltmarsh, sewage works, river islands and dunes. These are not counts so the percentages do not indicate the comparative importance of each category. They show where observers were accustomed to find nesting Lapwings and probably give a fair picture of habitat use over time.

The range of habitats used has changed little over the two centuries but Lapwings were reported less frequently in semi-natural habitats in the agricultural

Table 4.1. Habitat use by breeding Lapwings in Britain over the past two centuries, based on reports in regional and county avifaunas.

Period	*Semi-natural agricultural land*	*Improved farmland*	*Other*
19th century	58 reports (65%)	29 reports (33%)	2 reports (2%)
1914–1955	36 reports (50%)	30 reports (42%)	6 reports (8%)
1955–1975	25 reports (49%)	20 reports (39%)	6 reports (12%)
Post 1975	31 reports (33%)	44 reports (47%)	18 reports (19%)

sector and more frequently in improved farmland as the availability of these categories changed with agricultural improvement. A more important shift has been the frequency with which use of habitats outside agricultural land is now reported. Recent surveys support this, with 3.5% of the total population in such habitats in England and Wales in 1987, increasing to 8% in 1998 (Shrubb & Lack 1991, Wilson *et al.* 2001). In Wales 35% of the pairs found in 1998 were outside farmland.

POPULATION CHANGE

Figure 4.1 shows the main long-term population trends recorded in Britain, based on assessments published in county and regional avifaunas from the mid-19th

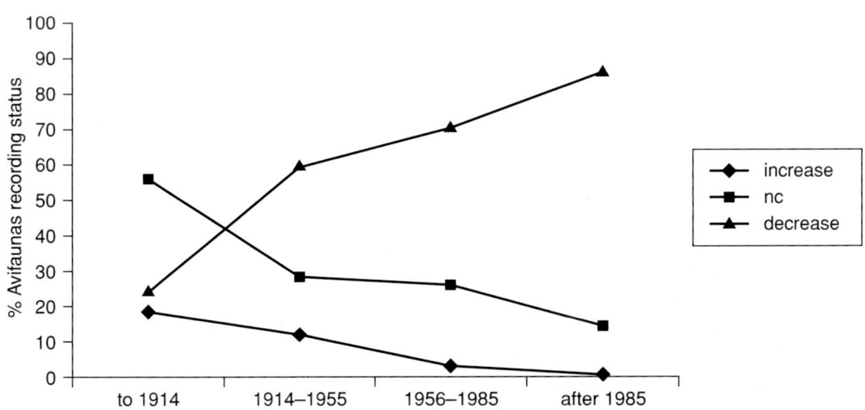

Figure 4.1. The percentage of county and regional avifaunas in Britain recording different changes in status in breeding populations of Lapwings. nc = no change recorded in the period. The period to 1914 covers the 19th century.

century onwards. Although enclosure, drainage and conversion to arable land in the species' core habitats in the first half of the 19th century brought about fundamental changes in the landscape, and marked declines in breeding numbers in some areas, particularly in the south and east, Lapwings eventually adapted to the mixed farming rotations which replaced their traditional habitats in the Waste. The 19th century avifaunas, most of which were published after 1875, then recorded a high level of stability in the population. Nevertheless a survey by the Board of Agriculture at the turn of the 20th century still showed marked decreases in 47 districts in England and Wales and 24 in Scotland, with increases in only 12 and five districts respectively (Collinge 1924–27). Since 1914 there has been an increasing tendency for county avifaunas to record declining populations although, locally, increases were recorded with the decline of egging following the 1926 Lapwing Act and with the introduction of three-year ley farming in the 1950s. There was a particularly sharp decline after 1939, when five million acres (2.02 million hectares) of grassland were ploughed in five years in England and Wales under wartime emergency regulations. Nest loss to these cultivations was widely regarded as responsible. Equally probable, however, was that productivity declined because ploughing grassland on this scale led to a widespread loss of good chick-rearing sites. The disruption caused by so large a habitat change in so short a time must have contributed. Since the 1960s the trend recorded virtually everywhere has been of decline and the species is nearly extinct today in Wales and southwest England.

Few historical reports are based on extensive censuses and some may reflect local rather than general change (Ticehurst 1932). In Figure 4.1 I have also had to assume that where accounts did not discuss population change it did not occur, which may not always have been true. Nor does the pattern shown by the county avifaunas always agree with the Common Birds Census (CBC, see page 47). Nevertheless the lack of unequivocal statements of increase in the avifaunas after the end of the 19th century outside northern Scotland and the Northern and Western Isles is striking. There seems little reason to doubt that the long-term decline in the British Lapwing population indicated in Figure 4.1 is genuine.

Other records support this. As part of the British Trust for Ornithology (BTO) 1987 Lapwing Survey of England and Wales, 27 areas, distributed across all regions, surveyed during 1956–1965 and then holding 637 pairs, were resurveyed. Numbers had declined significantly at every site save two and an overall decline of 61% was found (Shrubb & Lack 1991). The population estimate for England and Wales in 1987 was 123,134 pairs, so the decline measured in those 27 areas, if representative, suggests a population around 1960 of c.315,000 pairs. This is a considerably higher estimate than James Fisher's rough approximation (in Spencer 1953) of 350,000 individuals (say 175,000 pairs) in the 1930s. However, there may have been some population recovery in the immediate post-war period with the advent of three-year ley farming, which introduced a favourable, very widespread and close mix of spring cereals and grass leys into lowland farmland.

Only one Welsh site, holding eight pairs, was included in those 27 areas. However, Lovegrove *et al.* (1995) gathered data for breeding populations of

Lapwing from 31 Welsh sites censused during 1930–1959 and again during 1990–1993, which showed a decline from 360–374 pairs to 10–14 (-97%), and from 30 more sites censused during 1960–1979 and 1990–1993, which showed a decline from 291–318 pairs to 29–35 (-89%). Assuming their figures were representative, these authors estimated that a minimum of 15,000 pairs bred in Wales in 1970, and more earlier in the 20th century, and the decline there has probably been steeper than in England. Peers (2003) showed that the decline was particularly marked in Breconshire between the late 1970s and 1987. By 1998 the population in England and Wales had virtually halved to 63,000 pairs (Wilson *et al.* 2001), an overall decline since the early post-war period of c.80%. There was also a marked decline in range between 1987 and 1998, with 39% of tetrads visited occupied in 1987 but only 29% in 1998.

The BTO's Common Birds Census (CBC) shows a more complex pattern. Marchant *et al.* (1990) and Siriwardena *et al.* (1998) both illustrate the CBC Index showing marked general population decline starting around 1980. The figure in Marchant *et al.* also suggests that the population never recovered to more than around two-thirds of its 1962 level after the population collapse caused by the 1962/63 winter. The 1962 Index was based on rather few samples but there is no doubt that the value of CBC in monitoring Lapwing populations was severely distorted by that winter and the counts discussed above suggest that the pattern indicated by Marchant *et al.* is accurate.

O'Connor & Shrubb (1986) illustrated separate CBC Indices, which showed different patterns of change in cereal growing areas and in pastoral, mainly upland sheep farming, areas. In cereal-growing areas Lapwing populations declined slowly from 1962 to 1974 and then dropped sharply, but in pastoral areas there was recovery and increase from 1963–1980 and then a sharp decline. These patterns match changes in agricultural practice, with increasingly intensive cereal management throughout and a decline in spring cereals from the early 1970s, and a sharp increase in stocking densities in grassland during 1970–1990 (Figure 4.4).

In Scotland the Lapwing population has been more stable. Thom (1986) estimated a total of 75,000–100,000 pairs and O'Brien (1996) estimated 91,965 pairs in a survey of lowland Scotland in 1992 and 1993. A repeat survey of the same areas during 1997–2000 found 86,654 pairs, a decline of 6% (O'Brien *et al.* 2002). One difficulty with these surveys is the definition of 'lowland' adopted, which Wilson & Browne (1999) pointed out must have included considerable areas of marginal upland and upland, but the survey reports give no indication as to whether such land was actually surveyed. The terms 'farmed land' and 'lowland' used in the survey imply not. There is a potential source of over-estimation here which needs addressing and it should be noted that Wilson & Browne, using the BTO's survey methods, produced an estimate of 69,800 pairs, admittedly on a smaller sample of tetrads than planned.

There have also been local declines in Scotland of much greater magnitude than the 6% recorded nationally above. For example, on the Carse of Stirling Lapwings have declined by 94% since Galbraith's study in 1984–86 (Galbraith 1988a,

Sheldon *et al.* 2004) and a 76% decline was recorded on three areas of pastoral upland, totalling $20km^2$ in southern Scotland between 1980 and 2000, where the decline was particularly marked between 1980 and 1990 (Taylor & Grant 2004). The Breeding Bird Survey (BBS) has also recorded declines totalling 34% on 87 Scottish squares (O'Brien *et al.* 2002) and, more recently, declines of c.40% have been noted (M. Bolton pers. comm.). These figures indicate a more significant general decline. This was confirmed in upland areas by Sim *et al.* (2005), who found declines of 64–87% in half their Scottish study areas; a small increase of 12% was found in northeast Scotland.

Nevertheless, despite these apparent problems, Scotland has suffered less severe losses of breeding Lapwings than elsewhere in Britain and half or more of the British population is now found there. Scotland retains a much greater proportion of mixed farmland with spring tillage and a very much larger area of unimproved grassland, which comprises 74% of total agricultural land. Overall stocking rates and densities are also significantly lower there. All the recent surveys agree that overall breeding densities of Lapwings are two or three times higher in Scotland than in England and Wales, suggesting that habitat is more suitable as well as more extensive. Scotland also includes one of the most important breeding areas for waders in the northwest Palaearctic, in the crofting habitats of the machair in the Outer Hebrides. Here some 4,000–6,000 pairs of Lapwings breed on $186km^2$ of machair habitats, at a density of 20–27 pairs/km^2 (Fuller *et al.* 1986, O'Brien *et al.* 2002). Marked declines have occurred in this area in the past 25 years, largely as the result of nest predation by introduced Hedgehogs (Chapter 11).

SOME GENERAL ENVIRONMENTAL FACTORS

It is clear that Lapwings have declined at least regionally in Britain in periods of agricultural prosperity and expansion and in periods of agricultural decline. One important general point is that Enclosure, drainage and ploughing in its old core habitats during the 19th century led to the loss of large areas with a long history of stability and settled usage. Thus Jesse's description of the management of the sheep/arable system on the Sussex Downs in the early 20th century (Jesse 1960) would have been familiar to Cobbett riding over similar habitats on the Hampshire and Wiltshire Downs in the early 19th century and almost certainly to Defoe in the same area a century earlier (see Furbank *et al.* 1991). The grazing habitats of the Waste were widely managed under rights in common, which militated against change before Enclosure. For example, Moss (2001) noted that the economy and management of the Broadland grazing marshes changed little from the 13th century until well into the 20th. That may be an over-statement as there were drainage acts for ten parishes there between 1799 and 1807 (Ellis 1965) and, although sheep had been important in the 13th century, cattle predominated from the 16th. Nevertheless, long-term stability there must have resulted in stable food resources and nesting conditions for long periods of time and Moss also noted that grazing leases in the Broadland marshes limited both the period of grazing from 1 May to 24 October and the numbers of cattle per acre that could be grazed: stocking rates were very low by modern standards. All these factors would have favoured the large breeding populations of Lapwings and other wading birds. Beintema *et al.* (1985) showed that waders can adapt their breeding schedules to patterns of agricultural management.

The loss of these habitats had less immediate impact on farmland birds generally than might have been expected because they were initially replaced by mixed farmland of considerable diversity, offering favourable alternative habitats and resources on a large scale (Shrubb 2003). Lapwings certainly adapted, often by exploiting one habitat for nesting and another for feeding and rearing young. But successful adaptation to mixed farming rotations has inevitably exposed them to continued change in farming habitats and methods, from which they were largely insulated in the Waste and to which they are increasingly unable to adapt. The most obvious feature of the modern agricultural revolution has been the exchange of diversity which Lapwings were able to exploit for a steadily increasing pattern of uniformity in crop and management. The problems this presents in finding places to nest successfully and the time to do it affects not only Lapwings but also species such as Skylarks (Donald 2004).

A second general environmental factor of importance is the scale of land drainage over the past 200 years. We are accustomed to consider the arterial drainage of major wetlands such as the Fens and the drainage of wet grasslands as major habitat changes with marked impacts on wildlife. Less regarded but on an even greater scale has been the constant pressure of field drainage in improved farmland and, in modern times, the impact of water extraction on water tables and groundwater levels (Plate 16).

The practice of under-draining fields by placing drains beneath the surface has a long history, particularly in East Anglia (e.g. Pusey 1843), but it received a major boost in the early 19th century with the development of mass-produced clay drainage pipes (tiles) and the concept of thorough drainage – spacing pipe drains at uniform distances and depths across the entire field. Following the repeal of the Corn Laws in 1846 such work was also subsidised by cheap government loans. From about 1820 to the 1870s there was a massive surge in the under-drainage of fields, which various estimates have put at covering 8–12 million acres (3.2–4.9 million hectares) in England and Wales; in Scotland most arable land which needed it had been drained by the 1870s (Trafford 1970, Robinson 1986, Shrubb 2003). Nevertheless the impact of this work on wildlife appeared to be limited. It mainly concerned arable land and much old grassland was unaffected. Many 19th century drainage systems were poorly designed and deteriorated rapidly with agricultural recession from the 1870s, as the maintenance of outfalls and ditches upon which under-drainage depends was seriously neglected (Nicholson 1943, Sheail 1976, Shrubb 2003). Even so a survey by Belding (1971) found that 5.3 million acres (2.2 million hectares) of farmland in England and Wales were then still adequately served by 19th century drains. Drainage systems in Scotland were never so neglected.

From the beginning of the 1939–45 War a new campaign of land drainage started, with a major State input in organisation and finance. Because of this involvement we have good documentation of the scale of this drainage until 1980. The initial thrust of the wartime campaign was to repair ditch systems (Nicholson 1943) but throughout the period increasing areas of land were under-drained or redrained annually and, by 1980, the total had reached nearly two million hectares (Figure 4.2). Drainage has, of course, continued since 1980 but less information upon its extent is available with the withdrawal of state support. Nevertheless, Robinson & Armstrong (1988) estimated that 60,000ha annually were being drained in the mid-1980s. They also noted that Britain is one of the most extensively under-drained countries in Europe.

Modern drainage schemes are much more efficient than many installed in the 19th century, as systems are tailored for each individual field according to soil characteristics and so on. Puddles, pools and surface water, valuable for Lapwing chicks, no longer persist into the late spring/early summer, as fields now dry out much more quickly. Drinking water is also essential to Lapwing chicks (Chapter 12) and must be increasingly difficult to find in many fields. The impact of drainage on this scale is also cumulative, as lowering the water table in one part of a site affects water tables over the whole. Whole districts are becoming progressively drier, particularly in southeast Britain. Today this process is exacerbated by increasing water extraction for domestic and industrial use and for irrigation. The most notable symptom of what is happening is the loss of springs (Haslam 1991) and the drying-up of streams and rivers.

A clear indication of the importance of wetness to breeding Lapwings is the consistent finding recorded by surveys that the highest population densities occur

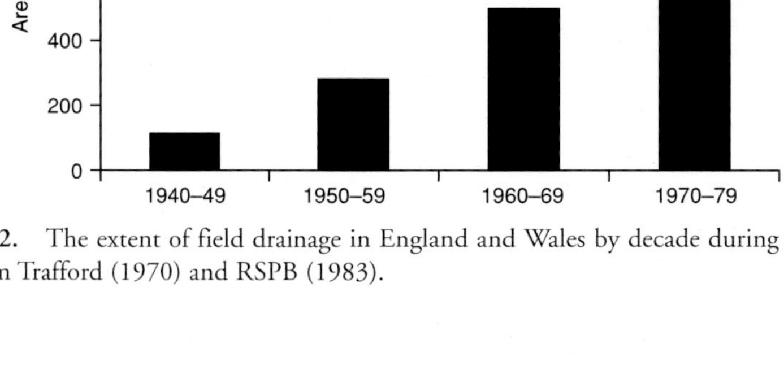

Figure 4.2. The extent of field drainage in England and Wales by decade during 1940–1980. Data from Trafford (1970) and RSPB (1983).

in wet grassland areas (p.60) and that very high densities can be established in wet grassland nature reserves (p.61). Wet areas are particularly valuable as feeding areas for breeding adults and their chicks and many studies have found a strong positive relationship between field occupancy and surface water (e.g. Berg 1993, Milsom *et al.* 2002, O'Brien 2002). Although puddles and surface water are important sources of surface invertebrate prey for chicks, Beintema *et al.* (1991) also showed that damp soils, from which they can obtain earthworms, may be essential in the later stages of chicks' growth. Lowering water tables makes such prey increasingly inaccessible. It also promotes more rapid and earlier vegetational growth and earlier crop management, factors which curtail the breeding season and reduce the chances of replacing lost nests (Beintema & Müskens 1987, Galbraith 1988a, Green & Robins 1993). Drought severely reduces breeding success (Jackson & Jackson 1980) and field drainage has been shown to be a major cause of Lapwings deserting nesting fields (e.g. Taylor & Grant 2004).

Lapwings can and do nest successfully on tilled land but rearing the young hatched depends largely upon the availability of suitable conveniently placed grass fields. Nevertheless, in Britain at least, when the habit of breeding in the mixed arable habitats of high farming emerged on a large scale, damp patches and puddles would still have been common features of both arable and grass fields, as I can attest from my own experience into the late 1950s. Matter (1982) showed that in tilled land in Switzerland, where chicks mainly foraged along the field margins, damp soils were significantly more productive feeding sites for them and they showed retarded growth under dry soil conditions. Fledging success also improved with the number of rainy days between 15 April and 30 June. Perhaps also increased rainfall provided a source of necessary drinking water in raindrops.

So far I have basically considered field drainage in lowland farmland but hill pastures have been extensively drained as well, both by pipe drains and by surface drains or grips. Such drainage is almost invariably followed by increased stocking rates, mainly of sheep, and the whole process has had significant impacts on the hydrology of many hill areas. Intensive high density grazing causes increased soil compaction through treading and increases the rate at which water drains off the hills (Carroll *et al.* 2004). It also affects the vegetation, particularly by converting dwarf shrub heathland to grass, a process which increases water loss by evaporation, as well as further encouraging run-off. Both processes mean that upland areas no longer hold water as once they did, to the detriment of upland birds, especially waders. It also exacerbates flooding problems downstream, which promotes demand for flood protection schemes detrimental to lowland wetlands.

There seems little reason to doubt that the dry conditions modern land management has increasingly imposed on the countryside have made an important contribution to the decline of breeding Lapwings by limiting nesting opportunities and, particularly, by making it increasingly difficult for them to rear the young they still hatch successfully.

NESTING IN ARABLE HABITATS

With the Lapwing's adaptation to high farming rotations in the 19th century a new core nesting habitat emerged. Mixed arable rotations came to provide secure nest sites on spring-tilled land and good fledging sites on grass, to which adults lead chicks after hatching. The proximity of tillage and grass strongly influenced chick survival: long journeys meant high losses (Galbraith 1988a). Easy proximity was typical in mixed farming rotations, which varied round a basic system of one or two years of leys, two or three of cereals and one of roots, and the three-year ley which followed, which comprised three years of cereals, usually spring barley, alternating with three years of grass.

Large scale adaptation to this habitat matrix probably took some time. Readiness to exploit tilled fields as nest sites may have been encouraged by the existence of a major population of Lapwings on arable land in the 19th century in the sheep/arable systems of the chalk and limestone, soils which were also among the earliest to be exploited by high farming (Grigg 1989). But successful breeding in the new arable habitats was probably circumscribed initially by the frequency of hand-weeding in both cereals and roots until quite late in the 19th century; quite apart from disturbance, eggs of ground-nesting birds were regularly taken to eat by weeding teams (Shrubb 2003).

However, surveys since the 1930s have consistently recorded a strong preference by Lapwings for spring tillage as nesting habitat in Britain. Table 4.2 shows habitat preferences for the main categories of agricultural land calculated from 11 studies. Habitat preferences were calculated as $100 \times \log O/E$, where O is the number of pairs observed nesting in each category and E the number expected if birds had distributed themselves in direct proportion to the area available. Such calculations can be misleading because pairs still nest in habitats indicated as 'avoided'. Table 4.2 therefore also shows overall densities as pairs/km^2 recorded in the 1987 and 1998 Lapwing surveys in England and Wales, confirming the importance of spring tillage.

These preferences are for nesting territories not overall breeding habitats but the table confirms that spring tillage has been a consistent preference for nesting over a long period: autumn tilled habitats have been equally consistently avoided. An interesting exception was recorded by Ammonds (1972) who noted that, in the 1930s, autumn wheat grazed off by sheep in the early spring, a not infrequent form of management then, also attracted many nesting Lapwings. Ley grass has also long been avoided, whilst selection of permanent grass habitats has been variable, as expected from their varied nature and management history. Rough grazing remains an important nesting habitat. These patterns have been consistent over time and on a country scale, a county scale and a farm scale. They also hold on a regional scale, as both Shrubb & Lack (1991, Table 4) and Wilson *et al.* (2001, Table 3) showed, and as has been observed in other individual studies (e.g. Galbraith 1988a, Taylor & Grant 2004).

Table 4.2. Habitat preferences of nesting Lapwings from surveys in Britain since the 1930s.

	Autumn tillage	*Spring tillage*	*Ley*	*Permanent grass*	*Rough grazing*	*Reference*
England & Wales						
1937	−25	+9	−25	+3	−1	Nicholson 1938/39
1960	−37	+17	−5	−11	+2	Lister 1964
1987	−59	+42	−19	−3	+19	Shrubb & Lack
	(0.28)	(2.86)	(0.71)	(1.00)	(1.67)	1991
1998	−84	+42	−37	−1	+23	Wilson *et al.* 2001
	(0.09)	(1.56)	(0.25)	(0.58)	(1.00)	
Scotland						
1981		+83[1]	−58[2]		+2	Galbraith & Furness 1983
1983	+8[3]		+1.5[2]		−4	Galbraith *et al.* 1984
1998	+8[3]		−1[2]		−3[4]	Wilson & Browne 1999
Cotswolds 1980	−18	+34	−203	+16	none	Knight *et al.* 1980
Sussex						
1983	−49	+22	−22	−45	+72	Shrubb 1985
Oakhurst Farm 1962–5	−11	+7	−30	+7	none	Shrubb 2003, Fig. 3.3
Oakhurst Farm 1975–8	−68	+68	−oo	+52	none	" "
Oakhurst Farm 1982–5	−110	+20	−53	+40	none	" "
Westmorland						
New Hall Farm 1964–71	None	+53	−7	−oo	−30	Robson & Williamson 1972

Notes. [1] tabulated as arable but text indicated spring tillage. [2] ley and improved permanent pasture not separated in Scotland, so improved grass. [3] recorded as arable but over 70% arable land in Scotland spring tillage at this period. [4] recorded as upland. Habitat preference is calculated as 100 × log O/E. Positive values indicate that the habitat was selected, negative values that it was generally avoided, -oo indicates that no birds were found. 'none' indicates that the habitat was not present. Figures in parentheses are densities in pairs/km^2.

These points suggest strongly that the decline of spring tillage in mixed farming rotations everywhere has contributed significantly to the Lapwing's recent decline as a breeding bird, as it has in Denmark (p.41). Spring tillage has declined by 60% in favour of autumn tillage since the early 1960s, affecting an area of c.2.5 million hectares (Figure 4.3). Shrubb (1990) showed a strong positive correlation between declining Lapwing populations in cereal growing counties in England and Wales and the declining area of spring cereals. Between 1963 and 1983 breeding populations in these counties, as measured by the CBC Index, fell by around 50% as cereals sown in spring fell from 75% of total cereals to c.20%.

What may now be more important is that the management of spring cereals has also changed. As in autumn, increased fertiliser applications and the use of pre-emergent herbicides accelerate growth and increase crop density, significantly

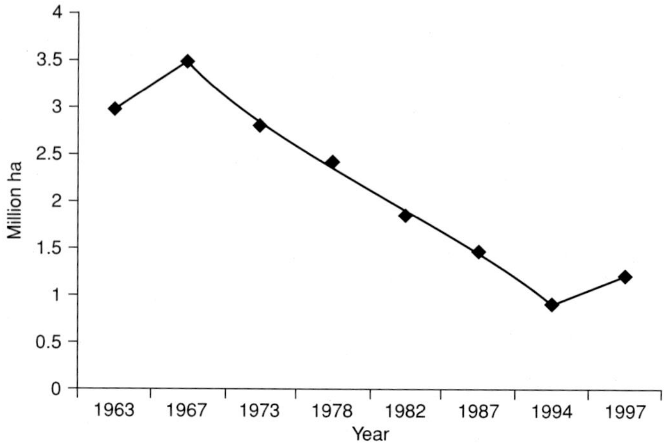

Figure 4.3. The decline of spring tillage in England and Wales from the early 1960s. Source: June Census statistics.

restricting the period when the habitat offers acceptable nesting conditions and reducing the chance of replacing lost clutches (Galbraith 1987, 1988a). Galbraith (1988a) also suggested that chicks also fared much worse in modern spring cereals than in the past. Spring cereals today are only nominally the same habitat as before the 1970s and they are declining in suitability as nesting habitat for Lapwings. Thus Milsom (2005) found that poor breeding success on spring-tilled land, nearly all cereals, rather than the declining area of such habitat, was the primary cause of decline of Lapwings breeding on arable land on his study area on the Hampshire Downs between 1981 and 1995. Also Henderson *et al.* (2004a) found that peas, sugar beet and set-aside fallows were now all more important nesting sites than spring cereals in the arable areas of East Anglia, East Yorkshire and north Lincolnshire because these crops were sufficiently low-growing at the peak of the incubation period (Plates 17–19). Set-aside fallows are also important for rearing chicks (p.167). As noted on p.42 Lapwings breeding in forage maize in Britain have experienced the same problems as those elsewhere in northwest Europe.

These changes have also coincided with the increasing separation of arable and pastoral enterprises, so that there has been an even greater loss of mixed farming habitats. Shrubb & Lack (1991) showed that in areas with much spring tillage but little grassland, Lapwings increasingly preferred grassland for nesting, but in areas where there was a moderate area of spring tillage in extensive grass, spring tillage was strongly preferred. This reflects the necessity for adjacent grassland for successful chick rearing. The polarisation of English farmland today between a pastoral west and an arable east means that these favourable conditions have diminished dramatically. Examination of the agricultural statistics for short term leys (under five years old) suggests that the availability of such habitat may have decreased by over 70% in arable regions since the early 1960s, rendering much remaining spring tillage of little value to Lapwings. Galbraith (1987) identified this as a particular problem on his central Scottish study area, where there has now been a sharp decline (p.47).

The main reason why Lapwings avoid autumn cereals so strongly today is the height and density of modern plant stands. These have been promoted by modern management techniques involving the use of autumn herbicides, which eliminate competitive weed growth at crop establishment, and very early applications of nitrogen fertilisers, which promote early and more rapid growth and tillering by the crop. The effect of both has been intensified by the increasing switch to earlier planting for autumn cereals, to September from October, which again promotes lusher plant stands earlier in the spring. Counts of nesting Lapwings on our family farm over 25 years showed that autumn cereals, although not a preferred habitat, were used regularly for nesting when the rotation brought them into the preferred nesting fields, until the new management practices were introduced in the early 1970s. Their use promoted more even plant stands and considerably increased cereal crop densities (by 27% between 1978 and 1984 for example). Densities of nesting Lapwings in the habitat fell from 6.5 pairs/km^2 up to 1965 to 3.5 pairs/km^2 during 1975–1980 and 1.5 pairs/km^2 in 1981–85, when nesting in the habitat had become irregular. Throughout the whole district of which the farm is part the impact of these management changes was to concentrate virtually the whole Lapwing nesting population of six farms totalling c.900ha, which once bred very largely on tilled land, into two areas of old permanent grassland totalling 60ha, at the extraordinary density of 90 pairs/km^2: both had areas of permanent standing water.

A more general example of the effect of these management changes was recorded by Shrubb (1985) in a survey of habitat use by breeding Lapwings in Sussex. In this survey 47.6km^2 of autumn cereals spread over the county was examined, of which 23.4km^2 were 'tramlined' (and therefore presumably intensively managed) and 24.2km^2 were not. Of 34 pairs recorded in this habitat only four (12%) were found in 'tramlined' crops but, in The Weald, less than 30% of autumn cereals examined were 'tramlined' and this crop was the principal nesting habitat recorded there. Poor drainage and heavy soils in this latter region often made autumn cultivations difficult, resulting in frequent bare or thin patches in the fields.

As noted in Chapter 3, the attraction that spring tillage often has for nesting Lapwings now involves the anomaly that breeding success is often very poor there. However, at least until the mid-1980s, hatching success there remained satisfactory because nests were secure from livestock and better camouflaged from predators. These factors significantly outweighed nest losses to arable cultivations or to tillage and, for example, 70% of nests in spring cereals were successful (Shrubb 1990). Females also lay larger eggs on arable land because of better food supplies early in the season (Chapter 10), which has considerable advantages and suggests that the early flush of food available on newly tilled land may partly explain this anomalous attraction.

NESTING IN GRASSLAND HABITATS

Until after the 1939–45 War there was rather little change in the management of permanent grassland, except that the numbers of cattle expanded at the expense of sheep. For Lapwings the most significant problem in grassland at this time was that farming abandoned significant areas of pasture (Shrubb 2003). Good examples can be found in Wales, where significant Lapwing populations bred in dune systems in Glamorgan, Carmarthenshire and Ceredigion when these were grazed by cattle (Heathcote *et al.* 1967, Salter 1895–1904). With the progressive abandonment of grazing since about 1940 Lapwings have deserted these sites. The decline of grazing management on the Gower, Pembrokeshire and Breconshire commons has had a similar result and Turner (1924) described the same phenomenon in the Norfolk Broadland marshes, as demand for the marsh hay declined with the demise of London's bus and cab horses.

Lapwings nesting in grassland select rather short open swards so some level of management, particularly grazing, is usually necessary. If available they favour areas where high water tables and/or poor soils delay growth in spring and maintain rather varied dun-coloured swards, which provide good cryptic defence for the nest (Plate 20). Such sites also tend to have low stocking rates and delayed spring turnout dates for livestock, reducing the risk of nest loss to trampling and desertion. The key grassland habitats are lowland wet grassland and the marginal grasslands at the moorland edge of the uplands, both of which exhibit these characteristics and attract high densities of nesting Lapwings (e.g. Smith 1983, Baines 1988), but pairs were also widespread until the 1960s in the unimproved damp pastures typical of much permanent grassland throughout farmland.

There has been a profound revolution in grassland management throughout pastoral farmland since the mid 1960s, involving extensive drainage, the reseeding of over 80% of enclosed permanent grass (Chalmers & Leech 1986), a decline in arable fodder crops, a switch from traditional hay crops to earlier and more frequently cut silage, large increases in fertilisation, particularly in the use of nitrogen, and marked increases in stocking rates and densities as sheep have replaced cattle (Vickery *et al.* 2001, Shrubb 2003). This revolution has led to significantly greater losses of some farmland birds in pastoral than in arable areas, particularly of species such as waders.

The most important changes have been drainage, reseeding and greatly increased levels of fertiliser usage, which have supported massive increases in stocking rates and densities (Figure 4.4) and more rapid and earlier spring growth of grass. The impact of drainage, increased fertiliser rates and reseeding was studied in detail in the English uplands by Baines (1988, 1989, 1990). He found that these processes resulted in significant declines in all breeding waders except Oystercatchers. Lapwing densities declined by 74% in improved pastures and 56% in improved meadows. Fewer improved fields were also occupied, with 90% of unimproved pastures supporting breeding waders compared to 44% of improved ones. The

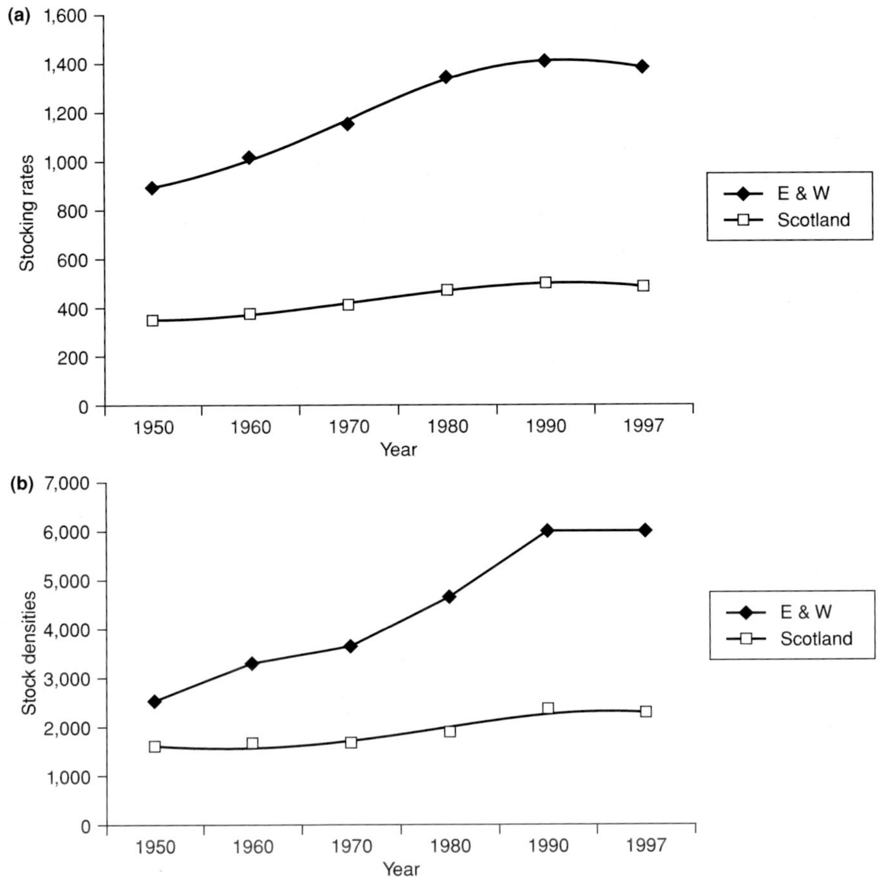

Figure 4.4. Overall stocking rates and densities in England and Wales (E & W) and Scotland at intervals since 1950. Stocking rates are livestock units per 1,000ha of all grass and fodder crops, densities animals per 1,000ha similarly. Livestock units equalise the differences between sizes and ages of animals in assessing grazing pressure. A dairy cow is one livestock unit, a beef cow 0.8, young cattle 0.5 and an adult sheep 0.15; see Coppock (1976) for a full definition.

declines in Lapwings were associated with sharp drops in productivity, which were largely the result of declining hatching success. Increasing numbers of clutches were lost to predation on improved pasture because reseeding altered the appearance and vegetational structure of the grasslands and nests became more vulnerable, a point Baines demonstrated experimentally. More clutches were also lost to farmwork such as rolling and harrowing on improved grass. Swindells (1997), discussing the disappearance of ground-nesting birds in Upper Nidderdale, also laid stress also on the general practice of chain harrowing in April and May and also on the shepherds' use of quad-bikes to do the rounds of the sheep, particularly during lambing in spring. He suggested that both factors are likely to cause extensive nest losses.

On commons in Breconshire, disturbance by scrambler motorbikes has led to marked loss of both Lapwings and Curlews (M. F. Peers pers. comm.).

Baines found that the impact of these losses was compounded by a sharp decline in the rate of replacement of lost first clutches, which he suggested resulted from the marked increase in stocking densities practiced, presumably as the result of site desertion (see below and p.153). Overall productivity declined from 0.86 fledged young per pair on unimproved grassland, sufficient for the maintenance of the population, to 0.25 on improved grass, well below that level. Interestingly, although Baines found that invertebrate populations were altered by grassland improvement, the changes were insufficient to affect fledging success.

There is no doubt that these factors have been general in grasslands in England and Wales and Shrubb (1990) concluded from an analysis of hatching success in the BTO's nest record cards up to 1985 that productivity was unlikely to be sufficient to maintain numbers in upland grasslands generally.

The increased stocking rates and densities shown in Figure 4.4 increase the risk of Lapwings in grassland losing nests to trampling, although it is uncertain how important sheep are in this context in Britain. Shrubb (1990) found a significant positive correlation between the incidence of trampling and rising cattle and sheep numbers in England and Wales. Other observers, however, have found that nest losses to trampling by sheep in upland areas tend to be low (Chapter 11). Sheep, however, cause other problems. They disrupt incubation, increasing the probability that nests will be destroyed by predators (see further discussion in Chapter 11) and, in Wales at least, they cause widespread desertion of nesting sites, particularly when ewes and lambs are suddenly turned into nesting fields which were empty of stock when the birds took up territories, which happens frequently (pers. obs).

The switch from hay to silage has affected 26% or more of all leys and permanent grass fields, rendering many of them unsuitable for breeding Lapwings, as growth is too lush in April for nesting and the early and frequent cutting may be destructive of both eggs and young, points which Lister commented upon as early as 1960 (Lister 1964). These factors are probably mainly important in dairying areas. Fields used for harvesting forage in livestock rearing areas in the uplands tend to be grazed until well into May and to be cut only once.

Taken together the factors outlined above amount to a major assault on the suitability of much grassland as a nesting habitat for Lapwings. Modern pastoral farmland has become an increasingly barren habitat for such birds. None of these problems is so severe in Scotland, where a very substantial area of unimproved grassland remains and stocking rates are generally lower than those in England and Wales (Figure 4.4).

Although decreasing for many years, the decline of grassland-nesting populations of Lapwings was sharply underlined by the 1998 Lapwing Survey of England and Wales. This showed that only in two regions, the North and Yorkshire and Humberside, was the loss of grassland nesting pairs less than 50% compared to 1987 (Figure 4.5). Major declines occurred in leys, in permanent grass and in rough

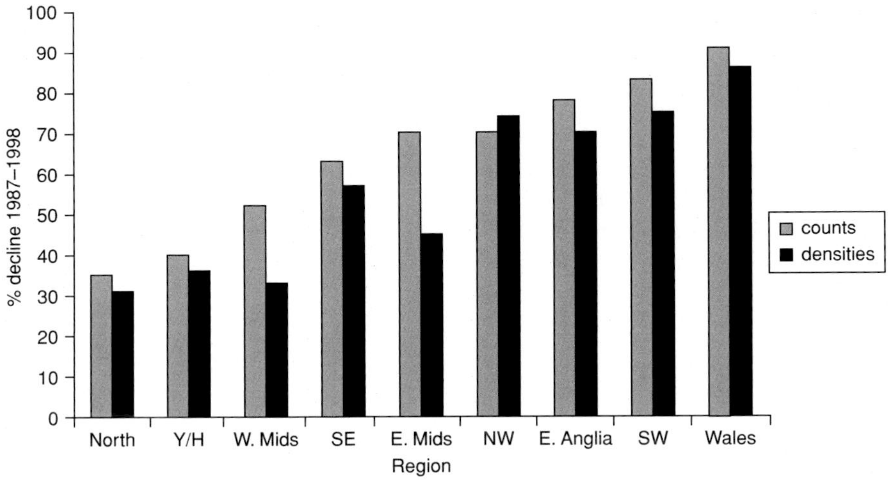

Figure 4.5. Percentage changes in breeding populations of Lapwings in grassland in England and Wales by region between 1987 and 1998. Regions are Standard Statistical Regions of England and Wales (as in Shrubb & Lack 1991). Y/H is the Yorkshire and Humberside region. Sources: Shrubb & Lack 1991, Wilson *et al.* 2001. The change in densities is shown as well as the direct counts because, although the same tetrads were visited in both surveys, the area of grassland available to the birds had frequently declined in 1998.

grazing as well as in each region. The only contrary trends were in the West Midlands and the North, where densities increased by 14% and 35% respectively in leys, probably the result of increasing concentration in fewer areas, and in the Southeast (numbers +55%, density +79% in rough grazings), which included some nature reserves or areas where habitat recreation work had been carried out (Wilson *et al.* 2004).

In Wales the population has collapsed. Here there were 166 pairs in 46,170ha of grassland of all categories in 1987, a low figure reflecting declines already in progress, but only 15 pairs in 33,124ha in 1998 when more pairs were found outside farmland. The opinion by Wilson *et al.* (2001) that Lapwings have always been comparatively scarce in Wales is refuted by evidence in Lovegrove *et al.* (1995) and by Holloway (1996), who showed the species to be common or abundant everywhere. Furthermore M. F. Peers (pers. comm.) found that breeding Lapwings were far more numerous in Breconshire and Radnorshire in the early 1970s than they were in Wiltshire. The species' greatest decline in Wales dates from after 1970, as do the major changes in Welsh agriculture (Shrubb *et al.* 1997). Distribution has also contracted sharply, particularly in southwest England and Wales, where declines have been greatest. Here 21% of tetrads visited held Lapwings in 1987 but only 9% did so in 1998 (Shrubb & Lack 1991, Wilson *et al.* 2001).

Lowland wet grassland has probably been the grassland habitat most vulnerable to change. Calculations by Shrubb (2003), based on Dargie (1993), Williams & Hall (1987) and RSPB (1983), suggest that an area of c.300,000ha of wet grassland was drained in England after 1970. In Wales Lovegrove *et al.* (1995) estimated another 100,000ha of damp pasture were drained up to 1992, two-thirds since 1970. Robinson & Armstrong (1988) noted that field drainage in Wales during 1970–1980 particularly aimed at dealing with waterlogging caused by springs and seeps, which were likely to be important Lapwing chick feeding areas. In some areas losses of wet grassland have continued over a longer period. Sheail & Mountford (1984) noted that 40% of Romney Marsh was drained before 1970 and most modern drainage in the Essex marshes occurred between 1940 and 1950 (Williams & Hall 1987). Highest rates of loss in eastern England have been to arable land, with losses of up to 4% per year in some districts (Dargie 1993). In the west Dargie noted that drainage was primarily for agricultural improvement of grassland. In this process fragmentation of blocks of habitat has also been marked. Although Dargie recorded 2,197 blocks of wet grassland, their median size was only 28ha: 1,024 blocks were smaller than 25ha. Fragmentation leads to declining species diversity, and lowering water tables in one part of an area increases the difficulty of maintaining water tables elsewhere there, again affecting wetland birds (see Williams *et al.* 1983, Green & Robins 1993). It may also reduce the population densities of Lapwings, reducing the efficiency of their anti-predator defence (p.161).

Successive surveys of waders breeding in lowland wet grasslands have shown sharp declines in these grassland populations of Lapwings. Smith (1983, revised by O'Brien & Smith 1992), in a survey of 1,282 lowland wet grassland sites totalling 240,000ha in England and Wales, found Lapwings present at 68%, with a total of 9,186 pairs. The survey was repeated in 2002, when 1,051 sites totalling 150,850ha were surveyed and 46% of sites were found to be occupied with 5,387 pairs reported. A decline of 37% was noted in the 851 sites which were covered in both surveys (Wilson *et al.* 2005). Although these counts only include some 7–9% of the total populations recorded in 1987 and 1998, the population densities indicated above were well above the general average shown in Table 4.2, suggesting clearly the importance of this habitat to Lapwings. An important finding from both surveys was the significance of a very few areas in these totals. In 1982 23% of the population was found on the Derwent Ings, Nene and Ouse Washes, North Kent Marshes and Somerset Levels. These sites held 30% in 2002, when the Norfolk Broads Environmentally Sensitive Area (ESA), the Avon Valley ESA and the Suffolk River Valleys ESA held another 16%. So large a proportion of the population concentrated in a few sites greatly increases the vulnerability of the species to continued habitat change and predation, although significant areas of these sites are now protected by nature reserves.

HABITAT USE OUTSIDE FARMLAND

Shrubb & Lack (1991, Table 4) found only 3.5% of the Lapwings in their survey nesting in habitats outside farmland. This proportion had increased to 8% in 1998, when overall numbers in such habitats had increased by 6.5% (Wilson *et al.* 2001, Table 3). The main habitats concerned were saltmarsh, dunes, gravel, sand and clay pits, wasteland, reservoirs, sewage works and airfields. In south Wales reclaimed opencast coal mining sites have been important in recent years (e.g. Dixon 1995), although these sites are vulnerable to redevelopment.

Analysis of BTO nest record cards from 1962–1985 showed that 18% of nests recorded were found in such habitats (Table 4.3). It is important to stress that these records are of nests, not pairs. Some may be repeat nestings and others were for successive years at the same site. Nevertheless a significant percentage of Lapwing nests are found outside farmland and a notable feature of the table is that 57% of the nests were located on land used for industrial purposes, habitat which is often undisturbed by human activity because of security concerns, despite the uproar which can go on around it. Lapwings appear to be adapting to a considerable scale of noise disturbance here and some even more eccentric nest sites have been recorded, such as roundabouts on busy main roads, flat factory roofs and gardens (Chapter 10).

Grassland nature reserves, many of which are managed by grazing, should also be considered here. Their area is growing, as is their importance for breeding waders such as Lapwings. All the major wet grasslands listed on p.60 include extensive nature reserve areas and other major grassland reserves for the species are Holkham NNR, Norfolk, with 240 pairs in 1996 (Taylor *et al.* 1999); Ynys-hir, Ceredigion, with 81 pairs in 2005 (R. Squires pers. comm.); Loch Gruinart, Islay, with up to 300 pairs during 1984–95 (Hampson *et al.* 1996) and Insh Marshes, Inverness, with 132 pairs during 1995–98 (Beaumont *et al.* 1998). Breeding

Table 4.3. Nesting habitats used by Lapwings outside farmland recorded in the BTO nest record cards during 1962–85.

Habitat	*Number of nests*	*Habitat*	*Number of nests*
Gravel pits, sand pits, quarries	152	Reed marsh and stubble	21
Wasteground & industrial sites	91	Waterfowl and country parks	16
Saltmarsh & shore	69	Reservoirs	14
Industrial and refuse tips	68	Dunes	12
Heathland	47	Golf courses	9
Sewage works	33	Fen, peat bog and peat diggings	9
Shingle and river islands	21	Other	44

At least 38 nests were also recorded from the islands off Wales but these then carried livestock and should perhaps qualify as farmland.

success is likely to be much better than in farmland, so that these grassland reserves will become a new core area for the species, whilst nature reserve status gives some hope of permanence for these populations. Lapwings have increased significantly in RSPB wet grassland reserves since 1987 and these sites support mean densities of 23.2 pairs/km^2 (Ausden & Hirons 2002).

There is also an important heathland population in the New Forest, Hampshire, which Tubbs (1986) estimated at 250–400 pairs in 1981. A fresh survey in 1993 found 104 pairs in 50–75% of the suitable habitat, suggesting some decline (Hampshire Bird Report).

CHAPTER FIVE

Distribution and populations in winter

The Lapwing is a summer visitor over most of its breeding range. Only in western Europe and the Mediterranean region do the breeding and wintering ranges overlap. In the east the winter distribution lies mainly between about 20°N and 35°N, from the Yangste Valley in China to Assam and northern Burma with a few to central Burma, northern Thailand, northern Laos and northern Vietnam. In India and Pakistan it winters in the Punjab and Sind east to Uttar Pradesh, with small numbers into northern Bihar and the Nepal lowlands. It is uncommon to locally common in Japan south of the island of Hokkaido and winters in southern Korea and occasionally in Taiwan and the Ryukyu Islands (Vaurie 1965, Ali & Ripley 1969, Sonobe & Robson 1982, Cramp & Simmons 1982, Smythies 1986, Roberts 1991, Grimmett *et al.* 1998, MacKinnon & Phillips 2000, Robson 2000).

Further west Lapwings winter in a number of sites in Afghanistan, Iran and the southern Caspian region, and in the Euphrates and Tigris Valleys into Turkey, where they winter extensively. They are winter visitors to northern Oman and now winter uncommonly over a substantial area of the Arabian Peninsula north of a line from Hufuf to Riyadh to Hali, but are scarce or rare in the Gulf States. In Israel Shirihai (1996) described them as very common or abundant, particularly in the north and west. Elsewhere around the Mediterranean, Lapwings winter in Syria, Lebanon, Jordan, Greece, the Balkans, Italy, Cyprus, Malta, Corsica and Sardinia, the Balearic Islands, the Iberian peninsula and along the North African littoral, with major wintering areas in northern Italy, the Nile Delta and Valley and northwest

Africa (Bannerman 1961, Vaurie 1965, Etchécopar and Hüe 1967, Hüe & Etchécopar 1970, Gallagher & Woodcock 1980, Bundy & Warr 1981, Jennings 1981, Bannerman & Bannerman 1983, Cramp & Simmons 1982, Hayman *et al.* 1986, Roberts 1991, Andrews 1995, Urban *et al.* 1996).

In western Europe the winter range lies mainly west of the 3°C January isotherm and entirely west of the 0°C isotherm (Imboden 1974), from western Denmark southwest to Spain and Portugal and northwest to Shetland and the Faroes. Lapwings are fairly common winter visitors to Madeira, the Canary Islands, where Bannerman reported that large flocks sometimes occurred on Tenerife, and to the Azores, where Bannerman noted that flocks of up to 70 had been recorded (Bannerman & Bannerman 1963, 1965, 1966; Clarke 2006). The northerly and easterly boundaries of the winter range in Europe vary with severity of the winter, with Lapwings shifting further west and south in severe winters. Thus, in mild winters, ringing recoveries show that some birds winter into north Germany, Norway and possibly Sweden (Imboden 1974), a pattern that now may be more regular with milder winters and fewer birds moving to the Mediterranean (Chapter 13).

This winter range is outlined in Figure 5.1. Over most of this range the northern and southern limits lie roughly along the January isotherms of 4°C (40°F) and 15.5°C (60°F) respectively, although in Arabia it now extends south to that of 21°C (70°F). The wintering range is also considerably smaller than the breeding range, presumably reflecting the Lapwing's strongly gregarious habits in winter, allowing

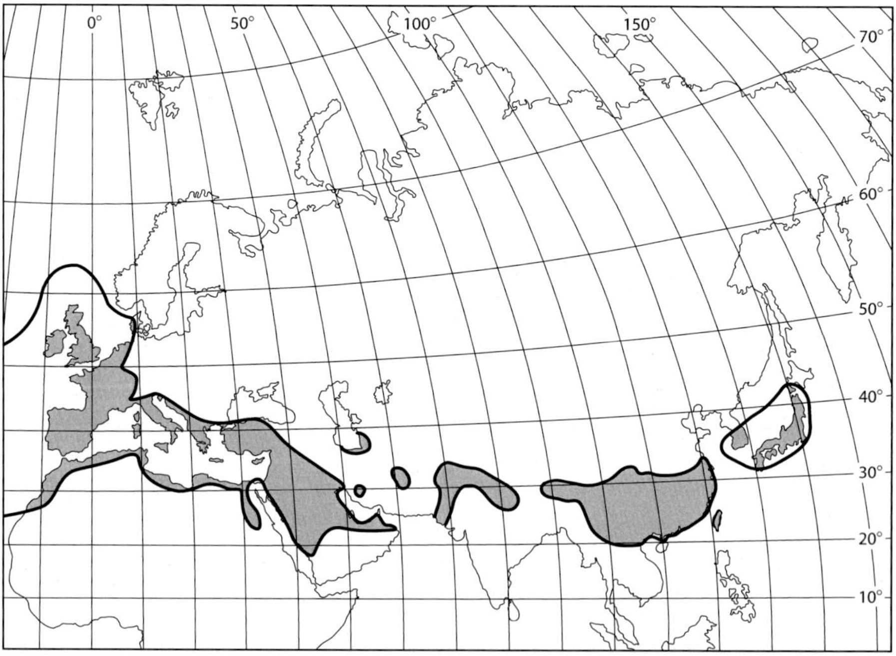

Figure 5.1. Approximate winter distribution of the Lapwing. See text for details.

large flocks to concentrate in limited areas of suitable habitat. It also involves the anomaly that a large migrant population can winter in a country such as Britain, which can only support a declining breeding population. This reflects the nature of the resources in farmland, the Lapwing's main breeding and winter habitat there. Crops are harvested and the great extent of bare or recently cultivated ground or young cereal crops means that soil and surface invertebrates, upon which Lapwings feed, tend to be most widely accessible if not most abundant in winter. Growth also declines with winter temperatures in grassland, increasing the accessibility of such prey there.

ABUNDANCE OVER THE WINTER RANGE

The most recent estimates suggest that some 2.8–4.0 million Lapwings winter in Europe and North Africa and a further 1.6–2.9 million in southwest Asia and the Caspian region (Wetlands International 2002). These estimates are derived from the breeding populations listed in Table 2.1, counted as individuals with an allowance of one juvenile per pair added (Stroud *et al.* 2004). Thus they may underestimate the European wintering population as they make no allowance for the presence of birds breeding further east, for example in west Siberia, which ringing records indicate as occurring in Europe (Chapter 13).

Wintering populations are very mobile, with birds shifting south and west in response to weather conditions, mainly declining winter temperatures. This influences distribution both within winters and between winters. Thus, in The Netherlands, atlas work in 1977/78–1982/83 found wintering birds in 86% of 1,760 5km-squares in November, declining to 61%, with a strong southwesterly bias, in January and increasing again to 84% in February (SOVON 1987). Actual population estimates were not made but extensive counts have been made in the Wadden Sea, where counts of wintering Lapwings have never exceeded 10,000 birds, with most normally found in the Dutch sector (Meltofte *et al.* 1994). The Netherlands is an important stopover region for Lapwings moving to wintering areas further west and south and a peak autumn count of around one million birds is thought likely to be typical (Chapter 13). The French Winter Atlas, covering the period 1978–81 (Yeatman-Berthelot 1991), offers a good example of the impact of mild or severe winters on the Lapwing's winter distribution in one area between winters (Figure 5.2). Kirby & Lack (1993) showed a very similar pattern for Britain and Ireland during 1981–84.

Within Europe Lack (1986) estimated that over 1.5 million Lapwings wintered in Britain during 1981/82–1983/84, Hutchinson (1989) estimated that 100,000–250,000 wintered in Ireland and Yeatman-Berthelot (1991) estimated that 1.6–2.0 million wintered in France during 1978–81, particularly west of a line from Aisne to Charente (see Figure 5.2a). These estimates were not based on detailed surveys but Trolliet (2005), in a very thorough census in January 2005 of

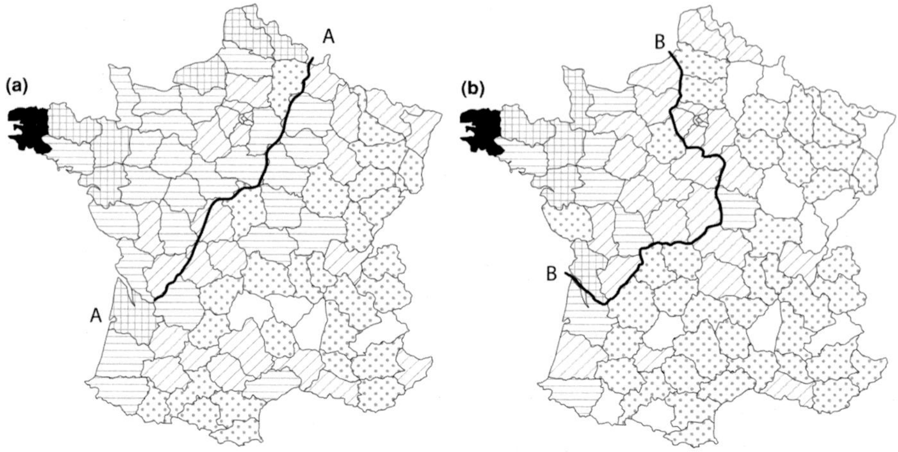

Figure 5.2. Midwinter distribution of Lapwings in France in a normal winter (a) and a severe one (b). Redrawn from Yeatman-Berthelot (1991). The line A–A is the Aisne-Charente line (see text). The line B–B encloses the Departments surveyed by Trolliet (2005), placed here for convenience; it does not refer to Yeatman-Berthelot's survey.

KEY

Number of occupied 20 × 27 km squares (Cartes I.G.N. at 1/50,000) per Department:

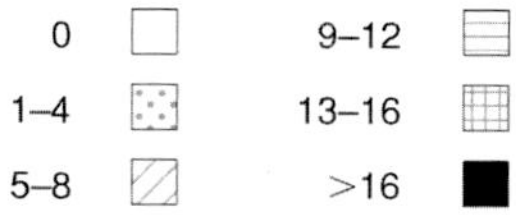

six regions of northwest France, Basse Normandie, Brittany, Centre, Haute Normandie, Pays de la Loire and Poitou-Charente (see Figure 5.2(b)) recorded a wintering total of 2.481 million birds (95% confidence limits 1.930 million to 3.100 million). Trolliet considered, from these results, that the Wetlands International figure of 2.8–4.0 million Lapwings wintering in Europe and North Africa was an underestimate. Clearly the total number wintering in France also considerably exceeds Yeatman-Berthelot's estimate, although the run of very mild winters since the early 1990s may have influenced this.

Lapwings winter throughout Iberia and Italy, although avoiding more mountainous regions (maps in Velasco & Alberto 1993 and Spagnesi & Serra 2003). Up to 2,000 winter in the Ebro Delta (Vilalta 1985) but elsewhere in Iberia it seems scarce, although the number of ringing recoveries there reported by Imboden (1974) gives some idea of scale. Finlayson (1992) described them as abundant in the south, with very large influxes during periods of cold weather in central Europe, for example of 250,000 between Jerez and the Marismas del Guadalquivir in November 1965. In Italy fairly substantial numbers (up to 80,000) winter in Lombardy (Fornasari *et al.* 1992) and flocks of 500–2,000 occur further west along the Po Valley in Piedmont (Cucco *et al.* 1996). In Sicily Lapwings occur mostly in tens but sometimes in hundreds (Iapichino & Massa 1989). Hundreds occur in some years in Cyprus but numbers fluctuate with winter conditions in Europe (Bannerman & Bannerman 1971). Other Mediterranean islands hold small numbers: for example up to 350 have been counted on airfields in Malta (Sultana *et al.* 1975) and a total of 1,000–2,200 recorded on Corsica (Thibault & Bonaccorsi 1999).

Lapwings winter in substantial numbers along the North African coast, although numbers again vary according to weather conditions in Europe (Urban *et al.* 1996). Thus on the Atlantic coast of Morocco up to 20,000 were present in the severe European winter of 1962/63 and 70,000–100,000 in January 1964 (Blondel & Blondel 1964, Thévenot *et al.* 2003). Numbers in the 1980s were more usually of the order of 7,500 and, in the mild 1990s, only some 3,000–4,000 have been present (Thévenot *et al.* 2003). They are common on the coastal plains of Algeria and abundant in some winters in northern Tunisia, but scarce in coastal Libya. In Egypt they are common in the Nile Delta, where flocks of several hundred are regularly noted, as well as at Wadi Natrun, Fayoum and along the Nile Valley to Asyut and Abu Tieg, where the winter population is estimated to be 500–5,000 birds (Summers *et al.* 1987, Goodman & Meininger 1989, Urban *et al.* 1996). A few may reach the extreme north of Sudan (Cave & MacDonald 1955). A few winter in desert oases in Algeria and, on the Atlantic coast, some penetrate south to the Senegal Delta (Urban *et al.* 1996).

Ringing recoveries show that wintering Lapwing populations in western Europe come from as far east as west Siberia (Dementiev & Gladkov 1969, Imboden 1974) but are mainly central and eastern European, Scandinavian and Baltic in origin, together with many from Belgium and The Netherlands. France and Iberia draw significant numbers from all these regions, although the patterns of recovery there may be biased by the influence of hunting (Chapter 13). Winter visitors to Britain come particularly from Belgium, The Netherlands and Scandinavia, while significant numbers of east European birds winter in Italy, mainly in the north. Nearly all these migrant populations reach down into North Africa. These movements are considered more fully in Chapter 13.

Further east 25,000–35,000 have been counted wintering in Iran (Summers *et al.* 1987) and they winter widely in Turkey— in Thrace, the southern and Black Sea coastlands, western Anatolia and the central plateau. Flocks of up to 4,000 birds indicate the presence of considerable numbers (OST Bird Report 1970–73). In Israel Shirihai (1996) estimated that c.100,000 were present in good (wet) years, with flocks of 500–2,000 and sometimes up to 5,000 not infrequent, particularly in the north. It is not uncommon in the Bekaa, Lebanon, and large winter flocks have been recorded on the Lebanese coast (Vere Benson 1970). It is still a widespread and fairly common winter visitor in the Syrian wetlands, with flocks of up to 2,300 at the Lake of Homs and flocks of 30–300 at 13 other sites in 2004 (Murdoch *et al.* 2005). It is a common winter visitor to agricultural land in Jordan, with flocks of 50–100 frequenting irrigated fields and smaller numbers on the agricultural fringes of the northern highlands and in the Azraq Reserve (Conder

1981, Andrews 1995). In Iraq it is a widespread winter visitor in small numbers on the Mosul plain in the north, with smaller numbers occurring in the centre and in the southern wetlands, where flocks of up to 70 have been noted (Moore 1956, Scott & Carp 1984). Information on numbers elsewhere in this region is scant. These birds presumably derive from populations breeding in the Middle East, the Caspian region and Siberia.

Many wintering Lapwing populations in the Middle East appear to be exploiting the modern trend to extensive irrigation, which is often converting desert to farmland, and this is probably leading to extensions of range, for example in the interior of Saudi Arabia. Here Rietkirk & Wacher (1996) noted that irrigation from deep aquifers has meant that available fresh water, green crops and tamarisk windbreaks are becoming a common feature in many parts of the Kingdom and dairy farming is important. They quoted the example of Thumamah in Central Province, where an irrigated dairy farm attracted flocks of up to 100 Lapwings, which they regarded as a fairly common winter visitor, mainly from mid-November to mid-February.

Little appears to be recorded about numbers wintering in the east of the range. Flock sizes in India and Pakistan appear to be small and both Ali & Ripley (1969) and Grimmett *et al.* (1998) noted Lapwings as often occurring in pairs. Roberts (1991) noted that it never occurred in the large flocks typical of birds wintering in Western Europe but noted one flock of 450. Wintering Lapwings were stated to occur in large flocks in China by Meyer de Schauensee (1984).

WINTER COUNTS IN BRITAIN

In Britain, although much information on wintering flocks is available in county bird reports, the only systematically gathered long term counts are those of the long series of estuary counts (now the Wetland Bird Survey or WeBS) since 1970/71. These show an interesting pattern. Both the January and the peak winter counts have increased by c.120–130% up to 2001/02. For example peak counts increased from an average of c.111,000 during 1970–74 to c.259,000 in 1998–2002 (Figure 5.3). The counts are shown as four-year means for convenience but both sets of counts show the same increasing pattern on an annual basis, although there was a marked trough, particularly in the January counts, during 1978–1987, a period of generally colder winters (Figure 5.4). The same pattern has occurred with Golden Plovers, which often share wintering grounds with Lapwings. County Bird Reports provide some evidence to show that Lapwing numbers inland in this period were higher than in the preceding or following periods (Chapter 6).

On an annual basis the WeBS counts also suggest that numbers in estuaries tended to stabilise during the 1990s, a point confirmed in a more detailed statistical analysis by Gillings *et al.* (2006).

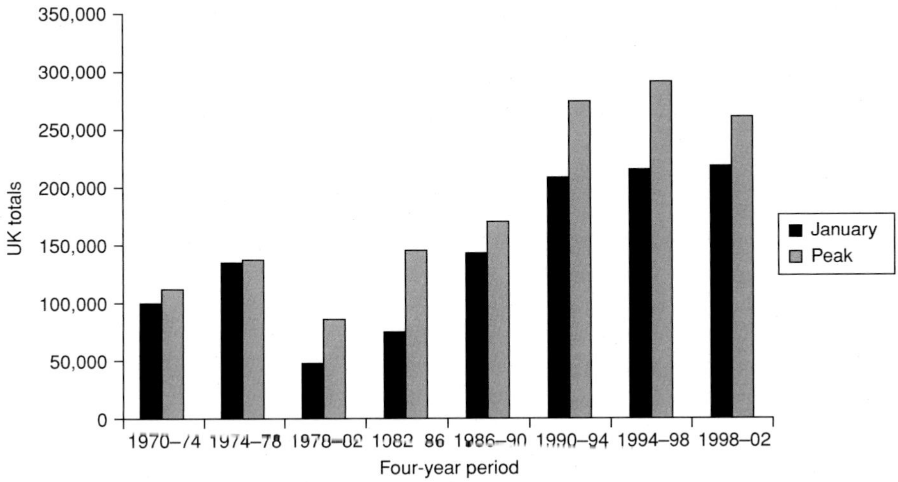

Figure 5.3. The increase of Lapwings recorded in UK Estuary Counts from October 1970 to March 2002. Figures plotted are the means for four-year periods for the January counts and of the peak counts in each winter.

How far Lapwings actually feed in estuaries out of preference on a regular basis, rather than under stress of factors such as severe cold, is not well understood. The main estuary counts are made at high tide, counting waders as they assemble at their roosts. The counts of Lapwings in Figure 5.3 therefore presumably mainly comprise flocks associated with high tide roosts of estuarine waders assembled on land adjacent to estuaries or using exposed sandbanks and islands as secure daytime

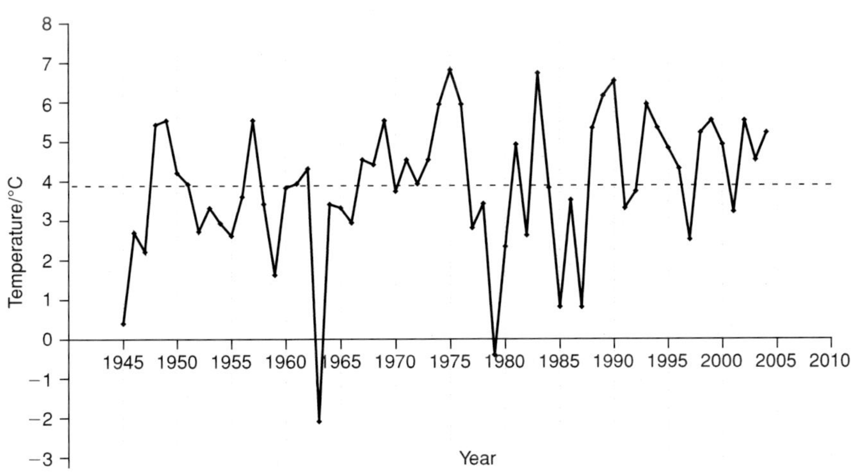

Figure 5.4. Mean January temperatures in central England from 1945–2004. The dashed line represents the overall mean for the period. Source: Hadley Centre for Climate Prediction and Research.

roosts. However low tide counts in the WeBS survey show that some flocks are now regularly found in estuaries (Chapter 6).

Gillings *et al.* (2006) analysed the coastal data in detail. They found that the most striking change has been a large increase on the east coast. Numbers along the south coast were more stable until the early to mid-1990s, when they also increased. Elsewhere numbers have declined. Thus coastal numbers have not only increased but there has also been a marked change in distribution, mainly to the east, which is discussed in Chapter 6.

The WeBS counts have also provided systematic figures from inland wetland sites since 1991/92, recording considerable numbers present, often in areas of wet grassland. It is important to stress that these areas are restricted geographically and hold only some of the Lapwings wintering inland. Nor is it clear whether these are feeding or roosting flocks or a mixture of both. Nevertheless Figure 5.5 summarises the published figures, showing the scale of numbers involved. These counts show the same broadly stable trend as do the main estuary counts during the 1990s, although numbers peaked in 1994/95. Although large numbers of Lapwings winter in farmland and other sites not covered by the WeBS counts, the broad pattern of stability the latter show at present may well apply to the British wintering population as a whole.

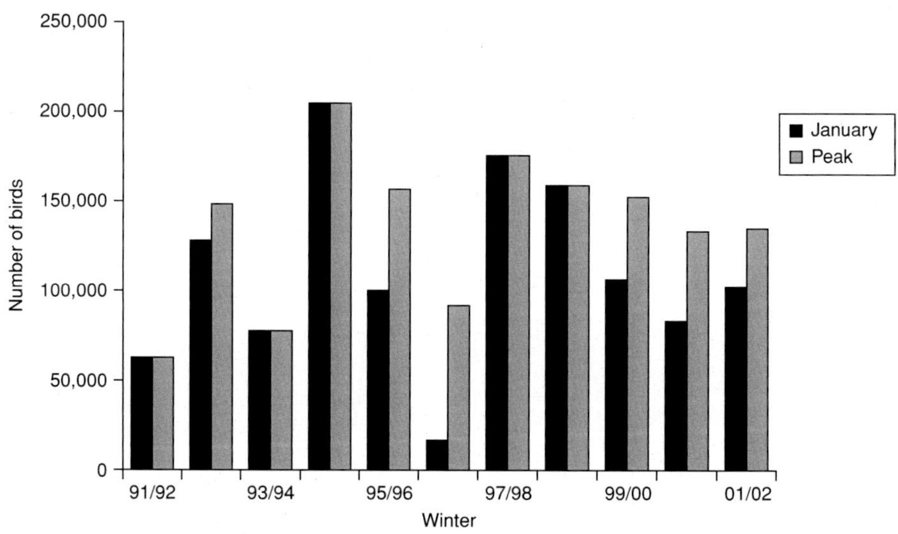

Figure 5.5. January and peak counts of Lapwings wintering at inland sites in Britain counted for WeBS during 1991/92 to 2001/02. Source: Wetland Bird Survey reports.

CHAPTER SIX

Winter habitat use

Much of the recent detailed work on the winter ecology of Lapwings appears to have been done in Britain and the account that follows is largely based on those studies. In farming terms winter can be defined as the period from late October to early March and I have used this definition here but migrant Lapwings arrive in Britain from late May (Chapter 13) and join native birds dispersing after breeding.

AUTUMN HABITAT USE

Late summer and early autumn is a difficult time for species which feed extensively on soil invertebrates. The growing hardness and dryness of the ground make such prey increasingly inaccessible and, in farmland today, most arable land during the period from June to September is occupied by standing crops which are usually too tall for Lapwings to feed in until harvest. Grassland is therefore the only extensive agricultural habitat then open to them and areas of damp grassland may be particularly important. Recent county avifaunas have shown that many Lapwings resort to wetland habitats outside farmland during this period, with sites such as reservoirs, gravel pits and sewage works being particularly sought.

In farmland in the past, however, rotations provided a series of summer feeding opportunities in arable land which the agricultural changes of the second half of the 20th century have removed. Before 1950 Lapwings were recorded feeding extensively in late summer and autumn in standing root crops, especially turnips, often on invertebrates associated with the crop, particularly the larvae of the Turnip Sawfly (now extinct) and the Turnip Moth. Harthan (1946) noted them feeding in large numbers on winter cabbage in the Vale of Evesham, taking slugs and other pests from the crop. Lapwings may not feed in such standing crops today at least partly because they lack such food sources in the pesticide age. Main crop turnips were sown around mid-summer, following a period of regular cultivations to control weeds, known as fallowing (Shrubb 2003). Thus their widespread inclusion in crop rotations also guaranteed a source of soil invertebrates in freshly turned soil in summer which, for Lapwings, was the most frequently used habitat in June/July in Nicholson's survey (Nicholson 1938/39). Payn (1962) noted their continued use of summer fallows in Suffolk into the early 1960s. Other standing crops were also used. Although they mainly used pasture, as late as the early 1980s I often found Lapwings on the family farm feeding or roosting in our clover leys, cut once for hay and left to regrow for a second crop of hay or for seed, during June to August. Again these birds apparently fed on the invertebrates associated with the crop.

Fresh feeding opportunities arise as crops are harvested, as harvesting often exposes a flush of food which birds exploit (O'Connor & Shrubb 1986). Mason & MacDonald (1999b), for example, found 30% of summer Lapwing flocks feeding in harvested potato fields and 11% in stubbles in Essex. In Sussex, burnt stubbles were avidly exploited by Lapwings and other birds from late July to early September. Harvested sugar beet fields and sometimes oil-seed rape stubbles are also exploited and, as the autumn progresses, arable cultivations become increasingly available. Although such food sources tend to be ephemeral, rotation and management provide a continuous series of opportunities over time.

GENERAL HABITAT USE IN WINTER

Lapwings presumably originally wintered in natural grasslands, heathland and marshland, probably in sites such as saltmarsh and the margins of water bodies and perhaps on estuarine mudflats. With the historical expansion of farmland in Europe (Chapter 3) wintering populations have become primarily birds of agricultural land. In northwest Europe Lapwings are traditionally regarded as grassland plovers, with lowland pastures as major habitats. In Britain historical information on the Lapwing's winter habitat is available from surveys in 1937 and 1960 and in the long series of 19th and 20th century county avifaunas. These records are summarised in Table 6.1. In compiling this table each record of use of each habitat was scored as one. No assumptions were made about habitat from names such as Romney Marsh, which can be misleading when much of the area has been drained and ploughed today. The records give no indication about the comparative importance of different habitats. Nevertheless it is clear that the broad pattern of habitat use has changed rather little over time, these records showing that Lapwings have long used a great variety of farmland habitats in winter, with arable habitats widely reported. Wintering flocks also regularly use habitats such as airfields, sewage farms, golf courses, playing fields and heathland. Heathland may have been an important winter habitat before enclosure in the 18th and 19th centuries, as it was for Golden Plovers (Shrubb 2003). Little indication of numbers was provided in the avifaunas but Nicholson showed that by far the largest flocks occurred in wet grassland and that flocks in arable habitats tended to be larger than those in other pastures. Lister (1964) recorded a similar range of habitats. His Table 3 shows higher average numbers in grassland than on arable land but his figures tend to show higher densities in the latter.

More recently in Britain a series of studies of wintering plovers in the late 1970s and 1980s, a period of generally rather cold winters (see below), found old pasture to be the most important feeding habitat for both Lapwings and Golden Plovers (Fuller & Youngman 1979, Fuller & Lloyd 1981, Barnard & Thompson 1985, Tucker 1992, Kirby & Fuller 1994). Nevertheless the winter distribution for Lapwings in 1981–1984 mapped by Lack (1986) showed little broad relationship with pastoral farmland, with many predominantly grassland counties in western England and Wales rather thinly occupied. Instead the core distribution lay in the

Table 6.1. Winter habitats recorded for Lapwings in Britain in the 19th century, early 20th century and after 1945.

	Number of accounts recording use of								
Survey period	*Marshes, wet grassland*	*Pasture and ley*	*Heath, common, Down, rough grazing*	*Fallow and stubble*	*Ploughed land*	*Arable crops*	*Estuary and mudflats*	*Saltmarsh*	*Other*
19th century	10	10	4	3	2	8	7	1	0
Early 20th century	19	24	12	14	10	9	4*	3	14**
Post 1945	17	13	2	0	0	16	11	1	13

Notes: *includes the margins of salt water; **includes aerodromes, playing fields, parks, golf courses, recreation grounds and sewage farms. Arable crops after 1950 includes arable land and subsumes plough and fallow.

Sources: county avifaunas and Nicholson 1938/39.

predominantly mixed farmland or arable farming counties of Oxfordshire, Buckinghamshire, Northamptonshire, Warwickshire, Worcestershire, Leicestershire and Nottinghamshire in the English Midlands, southwest into Wiltshire and east into Bedfordshire, Cambridgeshire and Lincolnshire (Figure 6.1). One characteristic of this core distribution was that it included the most extensive areas of open or common fields enclosed, mainly under Parliamentary Enclosures, in the 18th and early 19th centuries (Figure 6.2). After enclosure much of the common arable land on the clay soils in these Midland counties was converted to pasture (Hoskins 1955, Shrubb 2003), a practice which had a long tradition. Broad (1980) quoted good evidence that 'large areas of the southern and eastern Midlands, particularly Leicestershire, Rutland, Warwickshire, Northamptonshire, Buckinghamshire and

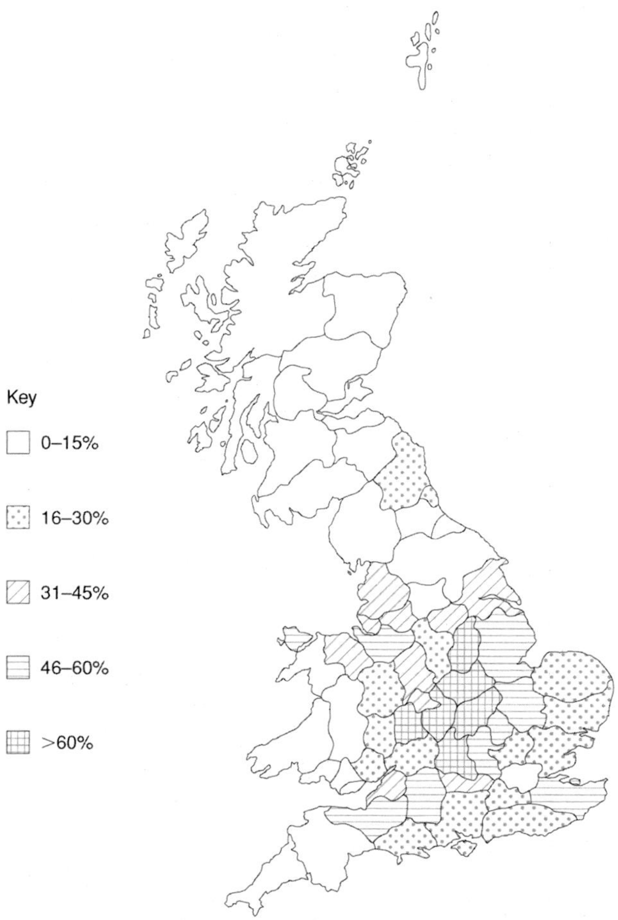

Figure 6.1. Core distribution of wintering Lapwings in Britain by county (region in Scotland) in the early 1980s. Shadings represent the percentage of 10-km squares in which the largest Winter Atlas counts (>1,500 birds) were made. Source: Lack 1986.

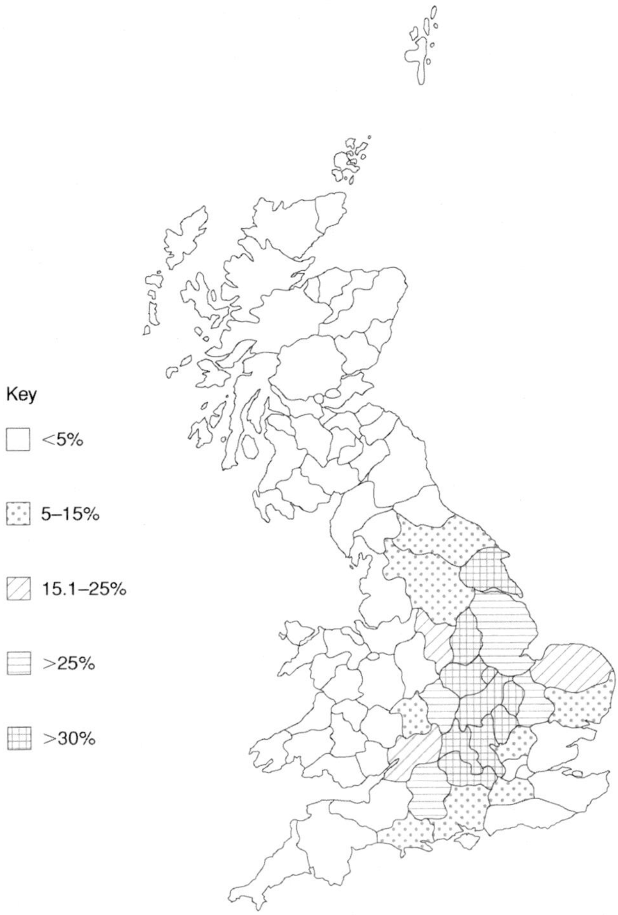

Figure 6.2. Percentage of agricultural area comprising open field arable land enclosed by Parliamentary Act during the 18th and early 19th centuries by county. Parliamentary Enclosure did not apply to Scotland. Based on pre-1974 county boundaries. Redrawn from Grigg (1989).

Bedfordshire, had evolved a system of specialized livestock farming on improved permanent pasture' by 1800.

A significant area of this pasture has not been ploughed since, evidence for which is provided by the archaeological feature known as ridge and furrow, the patterns of the old open field arable systems fossilised under grass. Harrison *et al.* (1965) and Sutton (1966) showed that this feature was concentrated in areas of heavier soils, generally avoiding the chalk and limestone of the Chilterns and Cotswolds for example. The Lapwing map in Lack (1986) suggests the same pattern. This pasture persisted because the heavy clay soils involved were unsuited for the rotations being developed in the 18th and 19th century until good underdrainage became possible on the clays around 1840 (Shrubb 2003). Farmers at this period also often found that levelling ridge and furrow on the heaviest clays left large quantities of intractable subsoils at the surface (Young 1813). It is only from the mid-20th century that many areas have been ploughed or, perhaps, that farmers have had the technical capacity to cultivate them readily (e.g. Chew 1953 for Leicestershire). Many of these grasslands are also highly productive and were long regarded as too valuable to plough. Those of Leicestershire and Northamptonshire were still classified as important livestock producing areas into the 1980s (Church 1975, Church & Leech 1983).

Some of this old pasture predates 18th century enclosure. Broad (1980), for example, discussed documentary evidence showing that arable land in the Vale of Aylesbury, Buckinghamshire, was being converted to pasture as early as the second half of the 17th century. Many of the parishes he discussed were included in an important unpublished study by R.J. Fuller of an area of some 400km^2 around and to the west of Aylesbury. He identified six main flock ranges in this area which supported a winter population of up to 20,000 Lapwings in the late 1970s and 1980s. These birds fed largely on pasture, large areas of which were in ridge and furrow giving a clear indication of its age.

Lapwings wintering in grassland feed mainly on earthworms. Several studies, for example Barnard & Thompson (1985) and Tucker (1992), have shown that the age of pastures is the main determinant of where Lapwings choose to feed because earthworm densities in pasture increase with age. Thus old pastures, at least theoretically, provide the most profitable grassland feeding sites and the core winter distribution of Lapwings shown in the Winter Atlas in the 1980s coincided with the most extensive remaining area of old lowland pasture in Britain. Two other areas with high wintering densities shown in Figure 6.1, Kent and Cheshire and neighbouring parts of the old county of Lancashire, also include important grazing (Kent) or dairying areas which have a long history of undisturbed grassland management.

Low tide counts in the WeBs surveys now show that quite substantial flocks regularly use estuaries in Britain (Table 6.2). Comparisons and trends cannot be calculated from these data as the counts cover different sites each year. It is also uncertain how far this is a new habit as these counts date only from the 1993/94 winter and the full results have only been published since 1997/98. However, Mason & MacDonald (1999a) noted an increase in estuarine feeding by Lapwings on the Essex coast, associated with a decline in the number using their farmland area nearby, in response to changing patterns of winter cereal growth. Elsewhere in Western Europe estuaries appear not to be significant feeding sites for Lapwings in Ireland, France or The Netherlands (Hutchinson 1989, Yeatman-Berthelot 1991, van der Winden *et al.* 1997). Sizeable counts have been made in the Wadden Sea area (e.g. Meltofte 1980) but these were estuary counts made at high tide roosts, not counts of feeding birds.

Detailed information on winter habitat use in other parts of the Lapwing's winter range is rather thin. Table 6.3 summarises the information given in standard

Table 6.2. Low tide counts of Lapwings on estuaries in Britain during the winters of 1997/98 to 2001/02.

Winter	*South coast*	*Wales & SW*	*Northwest*	*East*	*Scotland*
1997/98	3,766 (6)	1,348 (1)	15,706 (4)	8,283 (3)	2,525 (2)
1998/99	4,697 (5)	1,652 (1)	13,871 (3)	28,029 (6)	5,421 (3)
1999/00	2,577 (3)	959 (2)	5,451 (2)	4,047 (4)	6,025 (3)
2000/01	956 (1)	nc	nc	7,020 (4)	5,743 (2)
2001/02	3,141 (2)	968 (1)	4,943 (2)	10,777 (7)	1,597 (3)

Figures are the sum of mean counts for each site in each region, with the number of sites in parentheses. Not all sites were fully covered. Regions are South coast—Kent to Cornwall; Wales & SW—Wales and Bristol Channel; Northwest—Cheshire to the Scottish border; East—Thames to the Scottish border. Scotland includes the Solway Firth. Nc = no count. Source: Wetland Bird Survey reports. Note that the published counts do not separate feeding and loafing birds.

accounts. The list is not exhaustive but gives some indication of habitat use over a wide geographical area. Counts are lacking, so the relative importance of these habitats cannot be judged. Nevertheless the importance of farmland habitats everywhere seems clear, particularly as the category 'plains' here may well embrace extensive areas of agricultural land. The accounts also leave the impression that wetland and estuarine habitats may be more important than in Western Europe.

Table 6.3. Summary of winter habitats used by Lapwings in the Mediterranean region, North Africa and Asia.

Wetlands, marshes, margins of lakes and pools, riversides
Recorded for North Africa, Turkey, Israel, Syria, Armenia, Jordan, Iraq, Pakistan, India, Nepal, SE Asia.
Montane grassland
Recorded for Morocco.
Estuaries, mudflats, saltmarshes, saltpans and saltfields
Recorded for Morocco, Portugal, North Africa, Syria, China.
Wet grassland and grassland
Recorded for Israel, North Africa, Syria, Armenia, India, Pakistan, China.
Arable fields
Recorded for Morocco (plough, fallow and cereals) Sicily, Turkey (possibly), India & Pakistan (dry stubbles and fallows), Nepal, China (rice stubbles), Japan (rice stubbles), SE Asia.
Flooded arable and irrigated fields
Recorded for Israel, Jordan, Armenia, Saudi Arabia, India.
Coastal and other plains
Recorded for North Africa, Iraq.

Sources: Moore 1956, Ali & Ripley 1969, OST Bird Report 1970–73, Conder 1981, Sonobe & Robson 1982, Kersten & Smit 1984, Meyer de Schauensee 1984, Bijlsma *et al.* 1985, Inskipp & Inskipp 1985, Iapichino & Massa 1989, Roberts 1991, Andrews 1995, Rietkirk & Wacher 1996, Shirihai 1996, Urban *et al.* 1996, Adamian & Klem 1997, Grimmett *et al.* 1998, MacKinnon & Phillipps 2000, Robson 2000, Thévenot *et al.* 2003, Murdoch *et al.* 2005.

The expansion of irrigated farmland is providing important new areas of habitat in the Middle East (Chapter 5).

I find that winter Lapwing flocks are fairly quiet when feeding but flocks assembling or departing from a roost call a good deal, with what Spencer (1953) described as a conversational or peevish chorus of rather thin 'pe-wee' or 'peee-wi' notes, the first syllable rather drawn out, the second clipped or cut short, particularly as other birds come in to join the flock. Birds also use a variety of 'pee-wit' calls and the opening phrases of the song in these circumstances which rather supports Spencer's view that these calls have a spacing function, what he termed individual-distance calls.

THE EFFECT OF WINTER WEATHER ON HABITAT USE

The weather is one of the most important determinants of range and habitat use by Lapwings in winter and the main influence of winter weather on their behaviour has always been temperature. Lapwings primarily feed on soil invertebrates and detect their prey visually. Winter feeding habitats therefore need to balance the requirement for barer ground to detect prey easily and the effect of frost, which seals the ground and makes prey increasingly unobtainable; snowfall compounds such problems. Grass insulates the soil better than arable crops such as young cereals, allowing birds to continue using a familiar area in lower temperatures; old pastures are better insulators than young leys.

Lapwings adopt several strategies to cope with severe winter weather. If frost and/or snow are prolonged the birds move south or west into milder areas. Ireland, southwest France, Spain, the Mediterranean region or North Africa may all be important refuges in such conditions: indeed the Spanish name for the bird, Avefría, means 'bird of the cold'. However, if the cold snaps are of a short duration birds may not move far. Nevertheless extensive cold weather movements by Lapwings are well known phenomena (Chapter 13). Except in exceptionally severe and prolonged cold, such as in 1962/63, Lapwings return to their accustomed areas fairly quickly once conditions ease. Newton (2003) suggested that this avoided competition for food in hard weather refuges.

Recent studies in Britain have shown that Lapwings otherwise tend to switch between grassland and arable habitats for feeding in response to changes in winter temperatures. Such variations occur both within winters and between winters. Populations wintering on coastal farmland, where winter soil temperatures tend to be higher than further inland (Coppock 1976), often feed mainly on arable fields, particularly winter cereals (e.g. Shrubb 1988, Mason & MacDonald 1999b) and, in my experience on the southwest Sussex coast, have always done so. During 1983–86 I found that cereals following clover leys were the most strongly preferred feeding habitats throughout the winter but that the birds increasingly switched from feeding on other arable fields to old grassland from December onwards as temperatures declined (Shrubb 1988).

In the Vale of Aylesbury, Kirby and Fuller (1994) observed that Lapwings often selected tilled habitats for feeding in early winter, switching to grassland as the winter progressed (and temperatures declined). Timing varied between winters. In some years grassland was preferred as early as October, in others birds fed on tilled land into November, but grassland was always strongly preferred in December and January and was the preferred feeding habitat in cold spells. In February and March there was a tendency to return to feeding on tillage. Village & Westwood (1994) also noted a strong preference for feeding on tilled land and cereals in the autumn and early winter in the Welland Valley and the Cambridgeshire Fens, increasingly switching to pastures in December, when many Lapwings also left their study areas. A similar pattern of behaviour was observed in northeast France by Balança (1984), with Lapwings using grassland for feeding in cold periods and switching to winter crops in spring. Switching between local habitats also occurs with mainly grassland feeders. In Nottinghamshire, Barnard & Thompson (1985) noted that feeding on leys, which had less dense swards than permanent pastures, occurred mainly in mild weather. Such changes also occur during the day, for Kirby and Fuller noted that Lapwings fed on grassland more in the morning, when temperatures were generally low, than around midday and especially in late afternoon, when they tended to move to arable land.

Recently, besides the change in coastal distribution revealed by the WeBS counts discussed in Chapter 5, a significant shift in winter distribution in inland populations in Britain seems to have occurred. The Winter Farmland Bird Survey (WFBS) during 1999/2000–2002/2003 found proportionately more Lapwings wintering in eastern England than was the case in the 1980s (Gillings & Beaven 2004, Gillings *et al.* 2006). That this is a shift in the population rather than simply a change in relative abundance is shown by other changes. For example Lapwings (and Golden Plovers) have largely abandoned R.J. Fuller's Vale of Aylesbury study area since the early 1990s, although habitat change there has been limited and other grassland feeders, such as thrushes, starlings and corvids remain abundant (R.J. Fuller pers. comm.). Both plover species have also largely deserted the Nottinghamshire study area used by Barnard & Thompson (1985, Byrkjedal & Thompson 1998) and have deserted many interior parts of Wales (*Welsh Bird Report* 1990–2003, but see below). Gillings (2003) noted that very few large Lapwing flocks were found anywhere in western districts of Britain by WFBS counters and that many squares visited there held no Lapwings, although Lack (1986) found them to be widely distributed, albeit often in fairly small numbers.

Examining the county bird reports for the core counties in Figure 6.1 provides further information. Such data are often difficult to use because they are not gathered systematically. However the reports do adopt a practice of publishing details of the major winter flocks reported. This does not mean that all flocks were located but, over a decade, such records should firmly establish whether large flocks or smaller ones were the normal pattern. It seems most probable that this pattern provides a good index of the overall numbers of birds using these areas in each period and editorial comment in many of these reports makes clear that declining flock size was associated with a sharp decline in overall wintering numbers from the mid-1990s. Figure 6.3 presents the mean winter flocks sizes reported in each decade in seven core counties, Berkshire, Oxfordshire, Buckinghamshire, Leicestershire, Rutland, Northamptonshire and Warwickshire, plus Bedfordshire, based on 108 flocks in 1965–75, 327 in 1976–85, 508 in 1986–95 and 420 for the final period. Flock sizes virtually doubled during 1976–95 and then halved again, and this pattern was repeated in each county. Winter temperatures for each decade are also indicated but these averages do no more than indicate that winters during the middle two periods tended to be colder than the preceding or following, which is also supported by Figure 5.4. Nevertheless, temperature and flock size were inversely related. Gillings (2003) also showed that the period from 1979/80 to 1987/88 was characterised by very low minimum temperatures, many wet and snowy days and frequently frozen ground. In this central period the overall mean flock size recorded in these counties was 2,690 birds, 31% higher than the means shown in Figure 6.3 for 1976–85 and 1986–95. The Buckinghamshire Report included no counts before 1980 but Knight (1988) provided supporting evidence for the pattern illustrated in that county. Two county surveys found a total of 14,550 wintering birds in 1975 and 22,180 (+46%) in 1986. None of this proves that these birds have shifted eastwards but, as Gillings *et al.* (2006) point out, the pattern illustrated is the reverse of the coastal trend and is what would be predicted if inland birds stayed further east in milder winters.

Associated with this change in distribution, and presumably resulting from it, has been a much greater use of arable fields, particularly winter cereals, than grass for feeding (Gillings & Beaven 2004, Gillings *et al.* in press, *Ibis*). Eastern England has long been dominated by arable farming. This change in feeding habitats has probably also involved an important shift in the composition of the Lapwings' diet in winter, for earthworms are much scarcer in arable soils than permanent pasture,

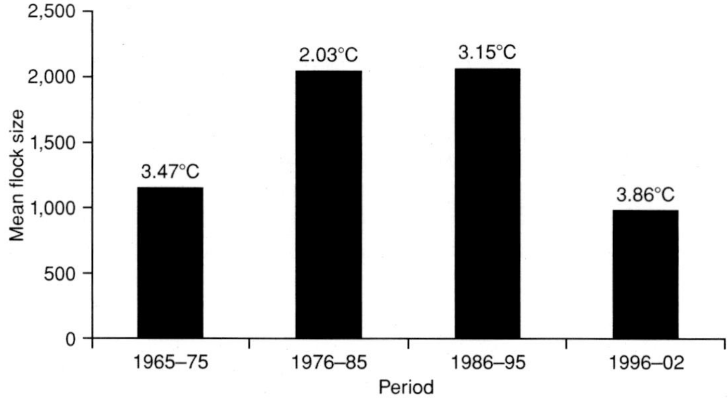

Figure 6.3. Mean wintering flock sizes of Lapwings recorded in four decades in core counties in Britain (see text and Figure 6.1) from 1965–2002. For sources see text. The numbers above each bar are the average of the mean minimum winter temperatures for each period.

so Lapwings feed to a much greater extent on smaller surface invertebrates which are less profitable (Chapter 7). In northwest France Trolliet (2004, 2005) found that about 60–65% of Lapwings counted in January 2004 and 2005 were also on arable land. In the absence of frost during the census periods, however, these birds fed much at night, often in different sites from those occupied by day, so that these counts could not fully assess habitat use.

It is uncertain why this shift in distribution has occurred. Chamberlain & Fuller (2000) showed that agricultural intensification between 1966 and 1988 led to considerable loss of grass to arable land in a zone extending from northeast England to The Wash and particularly through Midland England to the Severn, where grassland loss was most extensive in the core of the Lapwing's wintering range in the 1980s illustrated in Figure 6.1. So there have been significant habitat changes but these seem insufficient to cause a change in distribution. Lack & Ferguson (1993) observed that large areas of farmland always remained unused in Buckinghamshire and substantial areas of old pasture still remain in these core counties, totalling 315,000ha (28% of the agricultural area) (June Census Statistics 1997). Nor have Lapwings responded to grassland loss in their core area by shifting west into the nearby zone of predominantly pastoral farmland in western England and Wales, where the area of grassland has remained stable or increased. One reason may be that modern pastoral systems are often more intensive forms of land use than modern arable farming. They conspicuously lack the habitat diversity still visible in many arable systems. In pastoral farmland Lapwings undoubtedly seem to prefer old pastures and much intensively managed grassland is now regularly and constantly reseeded (e.g. Shrubb *et al.* 1997 for Wales). It lacks sward diversity and is heavily grazed, which may affect the diversity and abundance of invertebrates (Vickery *et al.* 2001). Nevertheless Atkinson *et al.* (2005) found that the intensification of grassland management may actually be beneficial to birds feeding on soil invertebrates in winter by enhancing populations of such prey. Evidence that intensively managed grassland is poorer habitat for wintering Lapwings than arable land is lacking. Overall it seems likely that the suggestion that this shift to the east has been the result of the modern trend to milder winters (Gillings 2003), is correct. Lapwings are making increasing use of arable habitats in winter as a result. This easterly shift has not occurred in isolation. Golden Plovers show very similar regional and coastal trends and many estuarine waders are remaining further east to winter rather than migrating to southwest Britain (Austin & Rehfisch 2005). That such distributional change is common to a range of wader species of differing ecologies and distribution at other times of year provides strong support for it being climate related.

This does not explain the anomaly of why wintering Lapwings do not exploit old pastures, with their high earthworm densities, at all times. One possibility is that Lapwings suffer significantly less kleptoparasitism in arable habitats (Chapter 7) and this partly compensates for any deficiency in food supply there. Their feeding rates in tillage or young cereals are often similar to or rather better than in old pastures (e.g. Gregory 1987, Shrubb 1988). Village & Westwood (1994) noted much lower feeding rates in permanent pastures than expected from the densities of earthworms present, probably because of the greater difficulty Lapwings had in detecting prey in the denser herbage. The accessibility of food, rather than its absolute abundance, may be more important to Lapwings and other farmland birds in selecting feeding sites (e.g. O'Connor & Shrubb 1986, Gillings 2003, Atkinson *et al.* 2004, Butler & Gillings 2004, Devereux *et al.* 2004). Atkinson *et al.* (2005) found a striking lack of any clear relationship between invertebrate prey abundance and field use in grassland by birds in their study. There was, however, a statistically significant tendency for birds to choose fields with short swards, although invertebrate abundance and richness tended to increase with sward complexity. Again this suggested that the accessibility of food was a critical factor in determining field use. Thus, with milder weather, arable habitats are preferred as offering easier living where food is more accessible on more exposed soil. There is no evidence, as yet, that modern arable habitats lack sufficient food for subsistence, although birds need to feed by night as well as by day for maintenance (Gillings 2003). This leads to the important point made here by Gillings, that site-based conservation measures may be ineffective for a species which periodically changes the range of sites occupied: sites which are unoccupied during a given period may be needed again in the future. The species' nocturnal behaviour (Chapter 7) complicates such matters still further.

Detailed examination of the present winter distribution in Wales shows an interesting contrast to the pattern described for England above, with an apparent westward shift towards the coast. Although estuary counts since 1988 have shown no clear trend in numbers, records in the *Welsh Bird Report* suggest that numbers wintering elsewhere in coastal districts have increased sharply since 1990, reversing a declining trend since the 1970s and 1980s. In contrast, numbers in inland counties have dwindled to almost nothing. Records from two counties show the pattern clearly. In Breconshire 1,000–4,000 wintered annually until the early 1980s (Peers & Shrubb 1990) but no more than 200 have been found in any of the past five years and only 54 in 2005/06 (M. F. Peers, pers. comm.). In Pembrokeshire Donovan & Rees (1994) recorded regular flocks of 200–400 in coastal fields which have now increased to regular flocks of 500–3,000. Records reported in the *Welsh Bird Report* show that 85% of all major winter Lapwing flocks (>500 birds) are now found within 5km of the sea and most of them are on the coast. It is not clear why this concentration has occurred but it seems likely to be habitat related.

SELECTING FEEDING SITES

Although it can be shown that Lapwings choose certain habitats for feeding, it is more difficult to say how they detect them or why they select particular fields. Barnard & Thompson (1985) suggested that in grassland they used simple visual clues such as local grass quality or the density of dung pats or molehills. In west

Wales I have noticed that birds often select fields at the lowest parts of the drainage catchment, suggesting that dampness is important. Because early in their breeding cycle Lapwings can apparently distinguish different types of grassland by colour (Klomp 1954), it seems reasonable that they can also detect the difference between old pastures and young leys from the density of the sward and turf. Such selection is less problematical in arable habitats, where cultivations in progress or crop growth send obvious signals.

Lapwing field and farm names can be found in several English counties (Field 1993), suggesting that tradition may be involved. Harthan (1946) recorded the existence of 'Peewit Lanes' in several parts of Worcestershire. Lapwing village names also occur, such as Tivetshall St. Mary and Tivetshall St. Margaret in Norfolk (Cocker & Mabey 2005). These clearly imply traditional sites, although perhaps mainly for breeding birds. Barnard & Thompson (1985) observed that wintering Lapwings and Golden Plovers had both been recorded using their study fields since at least the turn of the 20th century, although they did not say whether the record was continuous and the site has now largely been deserted (p.79). My own observations show that wintering Lapwings have used the fields of my Sussex study area continuously for over 50 years and they have certainly used them for longer, as my father often spoke of various fields on the site as particularly attractive to plover flocks. Fuller (1986) and Kirby & Fuller (1994) remarked on the continuity of use over winters in the Vale of Aylesbury, where there appeared to be traditional flock ranges. Nevertheless, within these, R.J. Fuller (pers. comm.) noted much variation in feeding distribution between winters. Nor did the flock ranges necessarily show long-term stability within the district. S. Gillings (pers. comm.) also commented that traditional patterns of use seem less obvious among Lapwings wintering in purely arable landscapes. Overall, the idea of fidelity to traditional areas or flock ranges is now undermined by the recently observed easterly shift in winter distribution.

There is probably an important exploratory element in locating feeding sites, perhaps particularly in arable areas. I found that feeding flocks in fields where they were only recorded once were much smaller than the overall average size for feeding flocks (mean 26 birds versus 55 overall). However, flock size increased if feeding persisted (Figure 6.4) so the continuing presence of feeding birds seems an important indicator of where to feed.

Other factors which have emerged as significant in the selection of feeding sites in recent British studies of wintering Lapwings are the age of pasture, the presence of livestock, field size, crop rotations and crop growth.

Age of pasture has been dealt with above but the presence of livestock has both direct and indirect effects. Grazing limits grass growth, enhancing visibility for feeding plovers. Milsom *et al.* (1998) found that feeding Lapwings avoided swards more than 13cm high and preferred those *c.*7cm high. Nevertheless Tucker (1992) found that, whilst there was a positive association between the presence of cattle and the presence of Lapwings, Lapwings avoided sheep, which produce much tighter and shorter swards. Buckingham *et al.* (2006) also noted reduced use of the

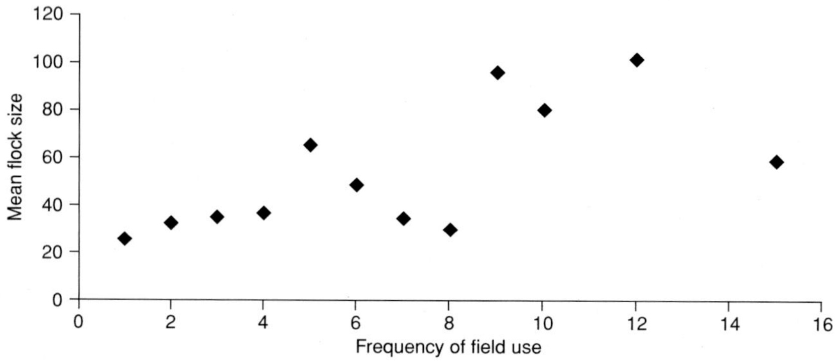

Figure 6.4. The flock sizes feeding in different fields on 873ha of mixed farmland in West Sussex in relation to the frequencies with which birds were recorded there. The relationship is statistically significant ($r_s = 0.69$, $n = 12$, $p < 0.05$).

shortest swards by species feeding on soil invertebrates and observed that such swards occurred on the most intensively grazed fields, where the compaction caused by the treading of livestock can have a markedly reduce the numbers of soil invertebrates. It seems likely that this is a particular problem in Wales and southwest England, where stocking densities of sheep are very high. Buckingham *et al.* also found that cattle were the preferred grazing animals for birds feeding on soil invertebrates, as the grazing and dunging patterns of cattle appear to promote the availability of soil invertebrates. Nevertheless, I watch Lapwings feeding in winter among quite high grazing densities of sheep in west Wales and it is possible that soil types are a significant element of these relationships, a point which would repay examination.

In arable habitats the presence of livestock influences rotations, being the basis of mixed systems of grass and arable crops. They thus have a direct and favourable impact on the structure of farmland habitats for Lapwings and other birds. I found that Lapwings selected cereal crops planted after a clover ley, where feeding rates were as good as in old pastures. This probably arose because the break in cultivations with the ley encouraged recovery in earthworm populations reduced by ploughing. Ploughing-under clover stubbles, where there is always significant regrowth, also adds much organic material to the soil: it is effectively a composting crop. The possible value of undersown cereals to wintering Lapwings seems worth noting, although such effects of crop rotation have not been noticed in Lapwings exploiting other arable systems. My own experience suggests that cereals following oil-seed rape may be another favoured rotation.

Livestock, particularly cattle, may also improve food supplies through the organic input from dung. Organic manure does not necessarily increase earthworm densities in grassland, but in the long term it will maintain them. Tucker (1992) found that the immediate advantage to Lapwings of spreading farmyard manure on grass was increased invertebrate activity near the surface, making prey more accessible. Tucker also noted that farmyard manure applied to arable fields, a declining practice today with the separation of arable and pastoral enterprises, did increase earthworm densities there, which increased their use by birds. Gillings (2003) made similar observations but considered it a minor influence on choice of feeding site by plovers in his study: but livestock was only a minor enterprise on his study area.

Some studies have found wintering Lapwing flocks to prefer larger fields (Crooks & Moxey 1966, Shrubb 1988, Tucker 1992, Mason & MacDonald 1999b). Gillings (2003) found that large fields were more often used than small ones but this was readily explicable by random settlement and did not, therefore, indicate a preference. How important any such relationship is to feeding birds is also unclear because not all studies have separated feeding and roosting flocks when considering it but I found that, apart from avoiding the smallest and most enclosed fields, feeding birds were indifferent to field size, whilst roosting birds always preferred large fields. I believe that this is a fairly general pattern. Nevertheless larger fields do attract larger flocks (e.g. Mason & MacDonald 1999b).

Barnard & Thompson (1985) also found that numbers at the best feeding sites remained rather constant. As numbers increased these sites were filled to capacity and birds spread into poorer feeding sites. If stable feeding numbers are characteristic of the best feeding sites, with territorial behaviour or other spacing mechanisms limiting access, increased field size is likely to confer the advantage of allowing more birds access because the favoured habitat will occur in larger blocks.

The evidence that types of field boundary significantly affect where Lapwing flocks feed seems somewhat contradictory. They often seem unimportant, perhaps because feeding birds are alert and difficult for predators to surprise. Thus Tucker (1992) and Mason & MacDonald (1999b) specifically examined this point and found it had no bearing on habitat or field selection. My Sussex site was very open but the distribution of the few large hedges or lines of trees also seemed unimportant to feeding birds. In the next parish, however, Milsom *et al.* (1998) found fields surrounded by hedges over two metres high and these were the least likely to be used by Lapwings. Gillings *et al.* (in press, *Ibis*) also found that field occupancy tended to decrease with increasing boundary enclosure. Nevertheless Plate 24 illustrates that proximity to buildings and substantial hedges does not bar feeding birds. As most modern studies have been done in a period when hedges have been widely removed, and many of the remainder are in poor or relict condition (Shrubb 2003), the patterns noted above may say more about the condition of modern hedges than the preferences of Lapwings. Farmland has been opened up by hedgerow removal throughout Europe (Chapter 3) and field boundaries may have had more bearing on winter behaviour in the past.

The largest percentage of wintering Lapwings feeding on arable land in Britain is found on winter cereals, now the most extensive crop habitat available. It is not necessarily the most preferred, birds often using cereals in proportion to their availability (Gillings 2003). In northwest France Trolliet (2004, 2005) also found that most Lapwings on arable land were on winter cereals, although they were not necessarily feeding there (p.81). There are, however, significant patterns of selection of more minor crop habitats in Britain (Table 6.4). These patterns of selection are based on preference indices, which showed that Lapwings used particular habitats in greater proportion than their availability in the landscape. It is interesting that oil-seed rape, after cereals the most extensive arable crop grown in Britain, is rarely used, although I found that cereals following rape crops were selected in autumn and Gillings found that Golden Plovers use oil-seed rape at night (p.89). Kallander (1977) found that rape fields were an important feeding site for Lapwings in spring in southern Sweden. He suggested that its growth pattern there provided an extremely favourable microclimate for earthworms at this season but this does not appear to be the case in Britain. Nevertheless the table again underlines the value of crop rotation and of the variety it brings in feeding habitats. Lapwings are very opportunistic feeders and quickly find and exploit transient food sources provided by cultivations and the harvesting of crops such as sugar beet. Harvested sugar beet fields are a particularly favoured habitat in eastern England. Many of the crop habitats listed in Table 6.4 are basically disturbed soil which produces, at least temporarily, a flush of invertebrates such as earthworms. Crop rotations and their management guarantee a succession of such openings and I suggest that opportunistic use of such temporary habitats within the cropping system is an important factor in enabling Lapwings to exploit arable land in winter.

Table 6.4. Selection of crop habitats by feeding Lapwings in Britain in winter. Grassland is included to illustrate seasonal shifts fully.

September	*October*	*November*	*December*	*January*	*February*	*March*	*Reference*
Pl	Pl	Pl	Pl	Ce	Gr	Gr	Mason &
Ce	Ce	Ra	Ce				MacDonald
Gr	Gr	Ss					1999b
	Cl	Cl	Cl	Cl	Cl	Cl	Shrubb
		Cr	Gr	Gr	Gr	Gr	1988
	Ws	Ab	Ab	Ab	Ab		Gillings
		Ss	Ss	Ss			2003
			Gr				
			Ss				WFBS from
			E				Gillings
			St				2003
			Gr				
			Ce				

Habitat categories are:- Pl—plough, Ce—cereals, Gr—grass, Ra—emerging rape crops, Ss—harvested sugar-beet fields (sugar-beet 'stubbles'), Cl—cereals after leys, Cr—cereals after rape, E—bare earth, St—stubbles, Ws—worked soil, Ab—arable break crops. Records from the Winter Farmland Bird Survey (WFBS) are in order of preference and are for the period from October to February.

Crop growth inhibits feeding by Lapwings. Gregory (1987), Village & Westwood (1994) and Mason & MacDonald (1999b) all noted that cereals higher than 7.5–10 cm were avoided for feeding, probably because taller crops increasingly obscure the ground, interfering with Lapwings' feeding strategy. In the second and third years of their study Mason & MacDonald (1999b) found fewer Lapwings using their site, perhaps switching to estuarine feeding, because crop growth exceeded the critical height by November. They ascribed this to more rapid crop growth resulting from warmer autumns but it may have derived from changes in the timing of sowing, which was becoming universal. In 1994 I found no Lapwings using the arable land of my Sussex study area. It was entirely oil-seed rape or cereals, the latter being unusually high and dense and completely obscuring the ground. This resulted from changes in the timing of cereal sowing, to September/October from October/November, and in drill technology, from drill spacings at seven inches (17.5cm) in the past to four inches (10cm) today (see Plates 26 & 27). Winter cereals undergo a period of dormancy in mid-winter, known as vernalisation and controlled by day length. So earlier sowing allows a longer period of growth before vernalisation and narrower drill widths compound the difficulties caused to feeding Lapwings. The significance of cereal sowing dates was neatly underlined for me in Sussex in the autumn of 2003, when drought precluded sowing before late October. In November/December the cereals there were once more at traditional heights and densities and my Lapwings had reoccupied their old feeding grounds.

However, Mason & MacDonald's conclusion that changed patterns of growth in cereals would pose problems for Lapwings wintering in arable land seems valid. So rotations that include crops such as sugar beet offer other advantages than favourable alternative feeding sites. Sugar beet is progressively harvested through the autumn and early winter and crops harvested up to early November are followed by winter cereals (Gill 1996). Thus the presence of sugar beet in a rotation means that later-sown cereal crops, which are more likely to be satisfactory feeding areas, will remain available.

ASSOCIATION WITH OTHER SPECIES

A number of field-feeding species, such as Starlings, occur in the same fields as wintering Lapwing flocks, and often with their flocks, but their presence is usually a matter of coincidence in a shared habitat. Golden Plovers, however, frequently associate with wintering Lapwing flocks, although the reverse is not necessarily true. Where studied, this association has proved to have several constant characteristics. Flocks comprising only Golden Plovers are infrequent by day, forming <1–10% of the flocks found in various studies, whilst Lapwing-only flocks comprised up to 70%. Golden Plovers also consistently use fewer fields in any area and associate with the larger flocks of Lapwings. They use these flocks to locate the best feeding sites and therefore parasitise Lapwings to this extent (Barnard & Thompson 1985, Mason & MacDonald 1999b, Gillings 2003, Trolliet 2004, 2005, Shrubb unpublished data).

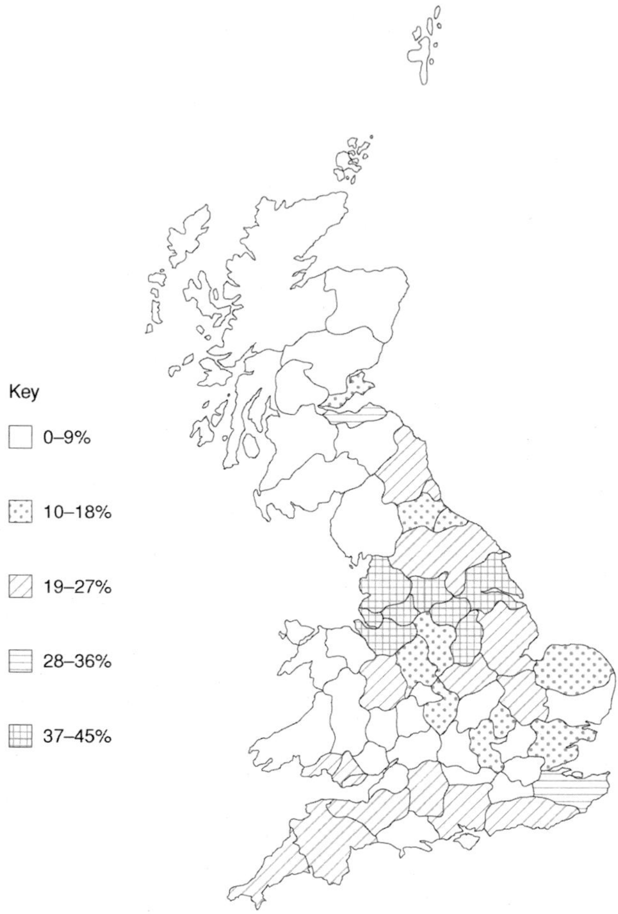

Figure 6.5. Core distribution of wintering Golden Plovers in Britain by county (region in Scotland) in the early 1980s. Shadings represent the percentage of 10km-squares in which the largest Winter Atlas counts (>496 birds) were made. Source: Lack (1986).

Nevertheless the winter distributions of Golden Plovers and Lapwings show some interesting differences. Figure 6.5 shows the distribution of Golden Plovers in the early 1980s mapped using the same criteria as that for Lapwings in Figure 6.1. On this basis the core distribution lay largely north of that of Lapwings and ran east to west across the country, including arable and pastoral counties. Overall the distribution so mapped showed no very obvious relationship to broad environmental features. Golden Plovers also appeared to be relatively more numerous in southwest Britain and more winter in Ireland than do Lapwings, with Lapwings more abundant in Britain (Kirby & Lack 1993). More recently both the WeBS counts and the WFBS have, as with Lapwings, recorded a very marked shift of Golden Plover distribution into eastern England, where the population now

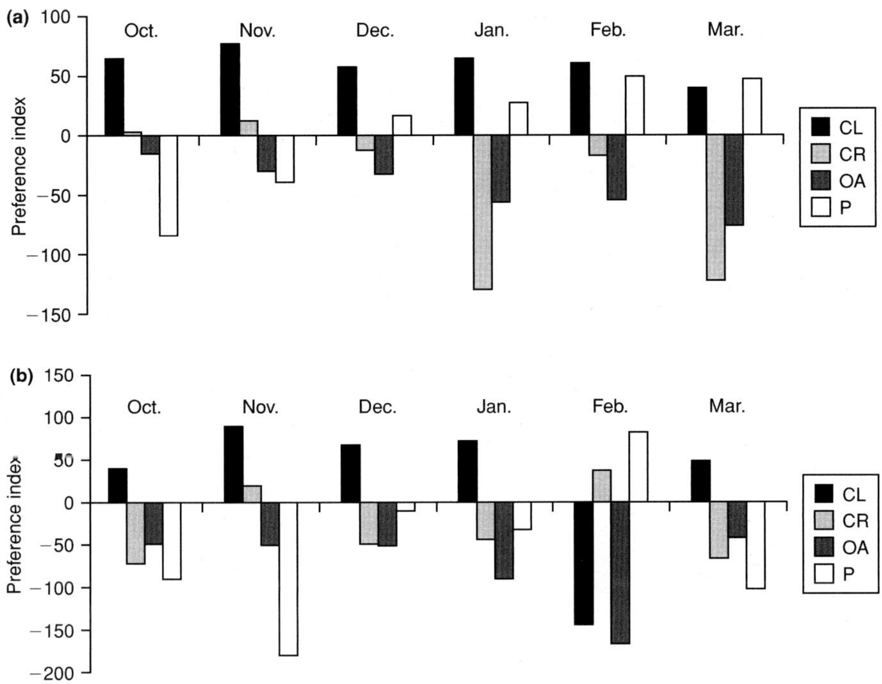

Figure 6.6. Habitat preferences of feeding Lapwings (a) and Golden Plovers (b) in winter on 873ha of mixed farmland in West Sussex during 1983–86. Preference calculated as log O/E × 100, where O is the number observed in the habitat and E the number expected if distribution was uniform in relation to area. Data from Shrubb (1988) and unpublished. Symbols: CL-cereals after leys, CR-cereals after oil-seed rape, OA-all other arable crops, P-pasture.

appears to be much more concentrated than that of Lapwings or than was the case in the 1980s (Gillings 2003, Gillings *et al.* 2006).

These differences in core distribution perhaps reflect differences in habitat choice between the two species. In two studies Fuller & Youngman (1979) found that, as with Lapwings, grass dominated habitat use by Golden Plovers in the south Midlands but Fuller & Lloyd (1981) showed that flocks wintering in East Anglia made much use of arable land. Figure 6.6 illustrates the habitat preferences of Lapwings and Golden Plovers in my Sussex study area. The differences were subtle but Golden Plovers made more use of cereals overall and avoided grass except in February, which tended to be the coldest winter month. Gregory (1987), Mason & MacDonald (1999b) and Gillings (2003) all found that winter cereals were the Golden Plovers' preferred and most widely used feeding habitat.

Gillings *et al.* (2005) also examined feeding behaviour at night. Golden Plovers then typically occurred in single-species flocks rather than with Lapwings, made much greater use of winter cereals than Lapwings and made significant use of oil-seed rape crops, which were ignored by both species by day. Thus there were marked differences in nocturnal ecology between the two species, which may influence their distribution. Gillings also noted that Golden Plovers may employ a different feeding technique at night.

The other species which habitually associate with winter Lapwing flocks are gulls, most usually Black-headed Gulls but also Common Gulls. They associate with them to steal food. We are so accustomed to the presence of gulls on inland fields and waters that it is often forgotten how comparatively recent their use of this habitat is. Hickling (1967) observed that the habit was first noted in Black-headed Gulls in the late 19th century and authors such as Yarrell (1845) made no reference to it, recording these gulls only as coastal birds in winter. Gilbert White, in *The Natural History of Selborne* (1789), does not mention the species at all.

Black-headed Gulls now winter widely in farmland, feeding on both grass and arable land. The habit of obtaining food by stealing it from plovers seems to have been observed almost as long as the gulls have been reported wintering inland. Laidlaw (1908) noted the habit as common in Scotland in the early 20th century but it was apparently not recorded there before 1894, or possibly 1888 (Evans 1908). Ussher & Warren (1900) also recorded the habit in Ireland at the end of the 19th century. As Black-headed Gulls also pirate food from shorebirds in estuarine habitats (references in Barnard & Thompson 1985), bringing the habit with them is perhaps unremarkable. One point about such piracy is that it is much more significant to the victim than the perpetrator. The details of this behaviour and its effect on Lapwings is discussed more fully in Chapter 7.

ROOSTING AND LOAFING

Wintering Lapwing flocks roost by day and by night. Generally, habitats used for roosting comprise ploughed or worked soil or tussocky grassland or similar terrain, where the broken nature of the ground provides shelter and concealment from predators. Against such broken backgrounds roosting flocks can be extraordinarily inconspicuous but the structure of these habitats interferes with the line of sight of feeding birds, making them poorer feeding grounds. Quite dense vegetation may be used. Baxter & Rintoul (1953) noted them as frequently roosting in winter in turnip crops in Scotland, a crop which M.F. Peers (pers. comm.) notes is still used in Breconshire, and, in autumn on our family farm, I often found them in stands of clover tall enough virtually to conceal them. Many studies of winter ecology in Lapwings have noted the difference between habitats used for roosting and feeding. Thus birds feeding primarily on pasture in the English Midlands roosted on tilled land (Barnard & Thompson 1985, Kirby & Fuller 1994). On my mixed farmland site in Sussex, where a high proportion of Lapwings fed on cereal fields, 57% of all roosting birds were found on grass, which was strongly preferred. Otherwise field size was the main determinant of where birds roosted. Where cereal fields were used for roosting there was a marked tendency to use early sown crops, which tended to be noticeably denser in mid-winter than later sown ones. Ploughed land was also used much more frequently for roosting than feeding (Shrubb 1988).

In areas dominated by winter cereals, where choice may be more limited, roosting commonly occurs in the same habitat as feeding (e.g. Gillings 2003) but Gillings still found a stronger selection for ploughed or bare tilled ground by roosting birds than feeding ones. Gillings also found differences between nocturnal roosting and feeding habitats. For roosting 52% of birds used harvested sugar beet fields, 20% bare tillage, 12% bean stubbles and 11% cereals, compared to 22% feeding in cereals, 23% in harvested sugar beet and 38% in bean stubble.

On tightly grazed sheep pastures in west Wales, where the option of using tilled land is rarely available, roosting Lapwings may select patches of ranker vegetation or mown soft rush or areas of molehills, which are sometimes very numerous and which again provide a broken background. Interestingly, in view of their hostile reaction to sheep on nesting grounds, I have watched sheep graze among a roosting flock without eliciting any obvious reaction.

Lapwings are also attracted to flooded pastures as roosts, roosting at the margins of the water which presumably also provides protection from predators. Similarly in a roost I watched in 2004/05 and 2005/06 at a small tidal lagoon in Meirionnydd the birds always roosted by day on the most recently exposed areas or in shallow water, never on the higher and drier areas of saltings, whatever the state of the tide. Near the coast I suspect that the use of estuaries may be increasing, contributing to the increase in the WeBS counts noted in Chapter 5. Thus in Sussex in 2003 I found that the Lapwings in my study area had deserted their traditional roosting habitats and were flying right out of their feeding range to roost on exposed banks and saltings in nearby Pagham Harbour, at least by day. Such a change in habit might arise because changes in crop management and rotations are reducing the availability of traditional or preferred roosting sites, such as plough or grassland, on farmland.

Roosting flocks have rather different characteristics from feeding ones. They are always more tightly packed, again presumably a defensive tactic. They are also larger. For example I found that the mean size of roosting flocks on my Sussex area was three times that of feeding flocks, as roosts absorbed birds feeding at several sites. I also found that roosting flocks of whatever size preferred large fields (Shrubb 1988) and Simon Gillings (pers. comm.) found the same in Norfolk, again presumably as a defence against predators. As with feeding birds (Chapter 7), I found that roosting flocks were highly aggregated, with many fields never used for roosting. Over the three years studied 75% of all roosting flocks were found on just nine fields and only 22% of the fields were ever used at all. Roosting flocks also showed something of the same pattern of variation as feeding flocks, as 58% of fields used for roosting were only used in one year. Although I detected no link between the choice of roosting field and the crop rotation there, this does suggest that some more subtle habitat factors operated.

As with nesting birds, some Lapwings, particularly in the industrial towns of northern England, have taken to roosting on the roofs of industrial buildings. This was first noted in Rochdale in 1984, possibly in 1972, and has now been recorded in Bolton, Bury, Oldham, Salford, Stockport, Wigan, Burnley and Darwen in Greater Manchester and Lancashire, Leeds and Bradford in Yorkshire, near Ilkeston in Derbyshire and at Washington, Birtley and Newton Aycliffe in Tyne & Wear. The habit has also been recorded in the southeast in Basingstoke in Hampshire, in Scotland at Cumbernauld and at Talke in Staffordshire (Calbrade *et al.* 2001).

These authors noted that birds were present on roof sites from early July to March, with most during September to February, and flocks of up to 200 birds counted on one roof. Roofs of quite steep pitch may be used and in some town centres a number of roofs have been used by the same flock over a long period. New roofs have been exploited quickly. As with Lapwings nesting in industrial sites, these birds have adapted to a considerable level of noise disturbance and human traffic. Calbrade *et al.* noted that this behaviour conferred the advantages of freedom from disturbance, security from predators, perhaps particularly foxes, and warmth from the buildings. In some sites the roofs are shared with Golden Plovers.

CHAPTER SEVEN

Food and feeding behaviour

Physically, like other plovers, Lapwings have rather short bills, ranging from 21.7–28.0mm in length (Dementiev & Gladkov 1969, Cramp & Simmons 1982). Plovers' bills lack the sensitive tip of many longer billed waders, which locate prey by touch. Instead the bill has a hard horny tip, well adapted to grasping and holding prey and perhaps to extracting food items from quite firm soil structures such as pastures. Byrkjedal & Thompson (1998) also noted that plovers' short straight bills could be swiftly aimed at prey which might be visible only briefly. As in other plovers, Lapwings also have noticeably large eyes relative to their head size. Although experimental evidence, quoted by Cramp & Simmons (1982), has shown that prey such as earthworms can be detected aurally, they very largely depend upon the visual detection of prey at or near the soil's surface, for which their large eyes are particularly well adapted. Rojas de Azuaje *et al.* (1993) compared the structure of the eye in the Grey Plover, which, like the Lapwing, forages by sight by day and night; the Greater Yellowlegs, which forages by sight during the day but by touch at night; and the Short-billed Dowitcher, which probes and locates prey by touch at all times. In the visual forager, the Grey Plover, the number of rods and the ratio of rods to cones in the retina were significantly higher. The rods in the retina are generally associated with good sensitivity to low light levels (Thomson 1964) so the structure of the retina was well adapted to nocturnal feeding by sight. It seems very likely that the eye of the Lapwing, which also feeds regularly at night, exhibits similar characteristics.

BASIC FEEDING BEHAVIOUR

As with most plovers, the main feeding actions of Lapwings are rather stereotyped—a few steps; pause; search or scan; move forward to stoop and take prey; trot on a few steps—a behaviour known as pause/travel feeding. Pause/travel feeding is thought to occur because prey detection is negatively correlated with speed of travel: many wader species are continuous visual foragers. Plovers may be pause/travel feeders because their eyesight is highly adapted to night vision and diurnal vision is compromised (Gillings 2003).

Most prey are taken from the soil but, in rough pastures, Lapwings also search grass tussocks for prey (pers. obs.) and, in the past, they picked it off standing crops such as turnips (p.72). I have also watched them in very dry autumn weather scratching over lumps of manure spread ready to plough-in, to get at invertebrates underneath. Spencer (1953) quoted two instances of Lapwings apparently taking insects in flight (although he was unconvinced) and they regularly follow the plough, at least in autumn, although tending to hang back from the mob of gulls and corvids similarly engaged. The 19th century literature includes quite frequent records of Lapwings, probably mainly injured birds, being kept as pets in gardens, where their utility in controlling slugs and snails was widely appreciated.

'Foot-trembling', pattering on the ground with one foot, has been widely observed by Lapwings feeding in grassland and intertidal areas. Although I have not personally recorded it on arable land, and Simon Gillings (pers. comm.) has found it to be very rare there, I find it happens frequently with birds feeding on grassland. Its function is uncertain but it may flush concealed prey or possibly encourage earthworms to move nearer the surface because it simulates rain. It certainly seems successful. In Wales, for example, birds watched feeding on a patch of wet grass at Ynys-hir were pattering with every scan. One leg was held forward and pattered swiftly and lightly on the ground; the birds often changed legs between scans. This was very effective at flushing earthworms and these birds were taking them, mostly small ones, at a rate of six per minute, always from a little distance in front of them.

Feeding Lapwing flocks present a characteristic appearance. On first arriving and starting to feed the birds slowly spread out until they are fairly evenly spaced. The course followed by feeding birds meanders as they tack about the field. Prey is often taken to one side with sharp turns and it may be taken more or less behind them. They then tend to continue in the new direction faced.

Detailed studies on arable land found that, on average, Lapwings moved seven or eight paces between scans (known as the interscan distance) and that 95% of all such movements were within 12 paces (*c.*130cm); 95% of prey captures were within seven paces, with larger prey taken at greater distances. Birds went to the edge of each scanned area on moving, so searching the entire area. Vegetation growth or the structure of the habitat may limit the range at which prey can be detected and interscan distances were greatest on cereal crops and on harrowed or bare soil, shorter on grass and shortest on ploughed land (Gillings 2003, Butler & Gillings 2004). Prey may be taken at once when detected or birds may crouch or stoop and watch more closely before striking. Barnard & Thompson (1985) noted that crouching was particularly aimed at targeting deeper and larger earthworms.

Lapwings feed by both day and night and the proportion doing so is related to moon phase and ground temperature (see below). High winds may limit feeding activity because the agitation of the vegetation obscures the ground more. Rainfall may be beneficial, as it encourages earthworm activity at the surface, but it may also influence feeding patterns by its effect on farm management, through varying the timing of cultivations for example. My experience in Sussex in the autumn of 2003 (p.87) provides an example. Frost influences the size and distribution of feeding flocks (see below).

FEEDING SUCCESS

Not all scans locate prey and not all strikes are successful. Gillings (2003), in arable land in Norfolk, noted averages of 3.9 scans per strike and 4.9 per capture, so that about three quarters of strikes were successful. In Sussex I recorded strike rates of 3.2 per minute in grassland and a mean of 2.9 per minute in cereals, where strike rates varied with the rotation. At least 88% of strikes were successful but the number of earthworms taken was quite low, not exceeding 18% of strikes (Figure 7.1): 19% of the birds watched took no earthworms.

Other studies of feeding Lapwings in arable land have found similar strike rates but success rates seem more variable. In particular Gregory (1987) recorded that 40% of strikes successfully obtained an earthworm in grass and 30% did so in cereals, whilst Gillings found that 50% of the individuals he followed to examine hunting success took no earthworms. Overall 85% of the prey he recorded by day were small invertebrates, mainly carabid and staphylinid beetles and millipedes, and 15% earthworms. He also found some element of prey selection, with significantly more large earthworms taken than expected from their frequency in the soil, although sampling problems may have influenced this. Carabid beetles tended to show the same pattern but the trend was not significant.

Most of the Lapwings I observed to take no earthworms were feeding in fields planted with cereals following oil-seed rape. Although I could not prove it, I was left with the strong impression that these birds were specifically hunting surface invertebrates, probably slugs in particular, which can be troublesome when cereals

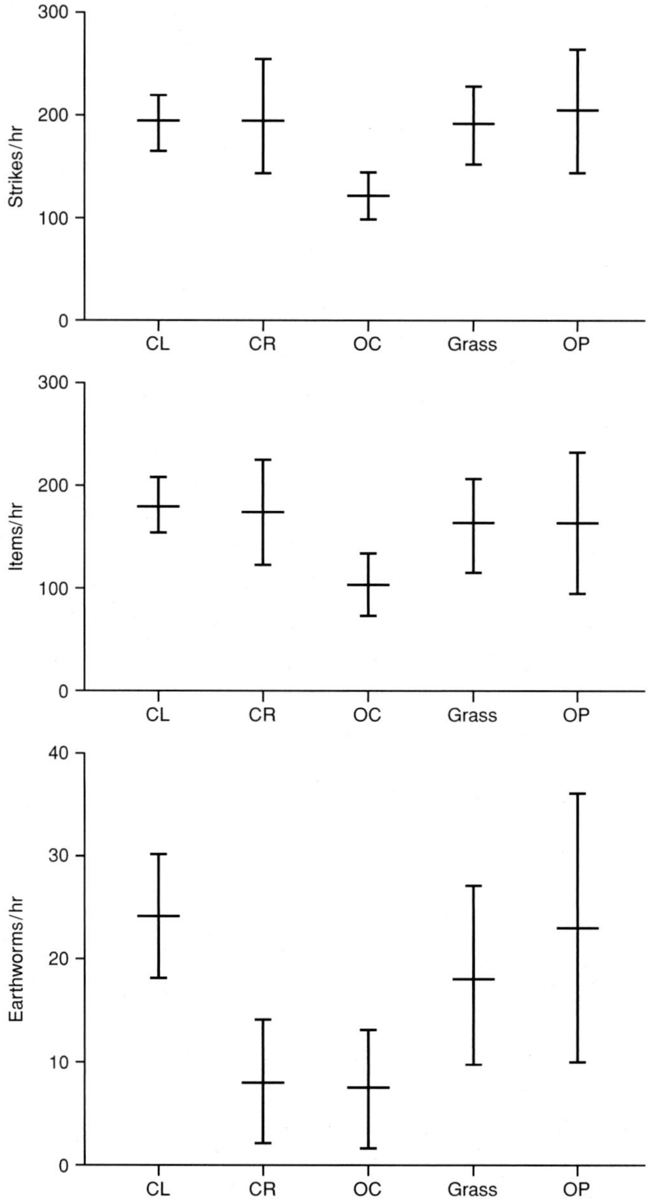

Figure 7.1. Feeding and success rates (means and 95% confidence limits) per hour by Lapwings feeding on 873ha of mixed farmland in West Sussex in winter. Habitats: CL – cereals after leys; CR – cereals after oil-seed rape; OC – all other cereals; Grass includes ley; OP – old pasture only. Data from Shrubb (1988) and unpublished.

follow a rape crop and have been widely recorded as Lapwing prey. Birds I watched feeding in an area of lightly grazed old pasture did not always seem to be hunting for earthworms either. Instead they carefully searched grass tussocks for surface invertebrates and this looked like a deliberate strategy. On sheep pasture in west Wales I have recorded much higher strike rates of 5.5 per minute but only 74% of strikes were successful. It was often difficult to assess prey taken but the numbers of earthworms definitely taken was lower than I expected at 27% of successful strikes and leather-jackets (cranefly larvae) may have been important on these reseeded leys, as Galbraith (1988a) found in Scotland. Högstedt (1974) found that 23% of strikes by his females successfully obtained an earthworm in the pre-laying season.

It is unclear how much such variations arise from the individual choices of the birds, which may be governed by other needs than straightforward energy intake, and how much from such factors as different soils and farming systems. Barnard & Thompson (1985) also found individual variations, with females more successful than males and adults more successful than young birds. Soils particularly may be important and, in my Sussex study, crop rotations certainly were. Feeding densities, success rates and the number of earthworms obtained all fell sharply as cereal rotations lengthened and cereals following leys were as good a source of earthworms as old pasture (Shrubb 1988. Figure 7.1). Such rotational effects are advantageous to Lapwings but they are not essential and Gillings (2003) observed that the diurnal food intake of Lapwings remains stable down to very low prey densities because their foraging method allows them to search large areas quickly.

DISPERSION AND TERRITORY IN WINTER FEEDING FLOCKS

Wintering Lapwing flocks tend to be highly aggregated within the landscape, with many of the fields in any broad habitat category in a particular area remaining unused. My Sussex study area comprised 111 fields, 18 in pasture and the remainder under arable crops, mainly cereals. Fewer than 40% of the fields were used in any year. The pattern of use changed annually. For example, half the fields used in 1984/85 were not used the previous year and 12 used in 1985/86 were not used in the previous two years; 39% of fields were not used at all, although some of these were used during a pilot study in 1982/83. Furthermore, not all fields were used equally intensively. Over the three winters nine fields attracted 45% of feeding flocks and 21 fields were used only once. Such aggregated patterns in the dispersion of feeding flocks are typical. For example, Gillings (2003) found a very similar pattern of occupancy in his arable area in Norfolk, with 45% of fields never used during the day in five years and, within a winter, only between 8% and 18% were occupied on more than 10% of visits. Other studies have produced similar results.

Variability between years in patterns of field use, such as occurred in my study area, seems fairly typical. Thus in extensive studies in the Vale of Aylesbury during 1981–89 R.J. Fuller (pers. comm.) found much shifting in the focus of feeding activity within one flock range between winters, with clusters or groups of fields several kilometres apart being used in different winters. Nevertheless two of these clusters were used in six and seven winters respectively, and one was used in four winters, so there was also continuity. Gillings noted a similar pattern of variability in field use in his arable area in south Norfolk.

In my study area the annual shifts in field use were consistently linked to the arable rotation which Lapwings followed, with a strong selection for cereals after leys and a secondary one for cereals after rape. There was a consistent decline in the use of cereal fields as runs of cereals lengthened. Pasture was used increasingly as the winter progressed (Shrubb 1988. Figure 6.6). Figure 7.1 indicates that these shifts were linked to food supply and availability. Seasonal variations in habitat use are also illustrated in Table 6.4 and again promote annual shifts in the fields used, as this is also linked to crop rotation involving annual changes in the cropping patterns of fields. As discussed below, other features of agricultural management affect invertebrate populations and may therefore also influence where birds feed and contribute to the volatility in feeding dispersion.

Flock sizes are highly variable. They vary with the persistence of feeding in a field (Figure 6.4 above) and with field size. They possibly also vary between broad habitat categories. Mean flock sizes given by Mason and MacDonald (1999b) and Gillings (2003) for their mainly arable sites in East Anglia were considerably larger than those I observed on mixed farmland in Sussex which, in turn, were larger than those recorded by Barnard & Thompson (1985) on pasture in Nottinghamshire. This perhaps simply reflected the fact that field size tends to be greater in arable farmland than pastoral. In northwest France Trolliet (2004, 2005) recorded a strong tendency for Lapwings to be distributed in rather small flocks with, for example, 35% of flocks in 2005 numbering 50 birds or fewer, a further 15% of 51–100 birds and about 10% of 101–150. Temperature also influences flock size as feeding birds spread out more in smaller groups at lower temperatures (Kirby & Fuller 1994, Gillings *et al.* in press, *Ibis.* pers. obs.). Nocturnal feeding flocks are also smaller and more widely dispersed than diurnal ones (see below).

Although birds appear to be fairly evenly spaced in feeding flocks, spacing often varies between flocks and within flocks. Gillings (2003) mapped flocks in which all the birds were feeding in arable land in East Anglia, which had a median density of 48 birds/ha (range 20 to 118). Densities in grassland appear to be considerably higher, Barnard & Thompson noting inter-neighbour distances down to as little as one metre. Kallander (1977) estimated distances of 2–5m between feeding birds in arable land in Sweden and I found similar distances in sheep pastures in west Wales: in frosty conditions individuals, like flocks, spread out much more. Flock density increases where more prey are available (Barnard

& Thompson 1985, Village & Westwood 1994, Shrubb unpublished data). Barnard & Thompson noted that increased density in feeding flocks reduced feeding efficiency. They remarked that this does not necessarily contradict the finding that flock density increases with the amount of available prey, as birds in such flocks may still do better than birds elsewhere, but it seems likely that this explains the pattern of birds spreading out more to feed in frosty conditions, when food is probably more difficult to obtain. More widely separated birds will interfere with each other less.

Gillings found that aggressive interactions were infrequent on his arable site, rarely occurring at densities of less than 200 birds/ha and densities below that level accounted for around 85% of feeding flocks. I also found aggressive interactions unusual in arable fields in Sussex but they are commoner in grassland as they increase with the density of feeding birds. Chases and fights often resulted in the loser flying forward in the flock and finding an open space in which to settle. Individuals may defend feeding territories within a flock and Barnard & Thompson noted that males were four times more likely to do so than females, with territories ranging in size from $2m^2$ (in patchy snow) to *c.*$80m^2$. Territorial disputes may be settled by overt aggression or through more formalised displays. The latter are very similar to boundary disputes on breeding territories. Opposing birds may face each other in the high upright pose (Chapter 9) with the head up and bill pointing down, tail depressed and wings held a little way from the body, and display the breast band. They also do parallel walks along the presumed boundary in this pose. More frequently they flex their legs into a half crouch with the head slightly pulled back and bill held straight out and raise the tail, sometimes turning to display the undertail coverts. In more intense displays the back feathers are raised or ruffled and the crest is raised and birds will then run at their opponent in this pose. They will also open their wings and flash their wing linings at opponents. If these displays are unsuccessful the birds fight, buffeting each other with their wings and occasionally fluttering up, as they do in the breeding season (Figure 7.2).

Although he entitled it pre-mating courtship Brown (1926) accurately described such territorial behaviour 70 years ago, noting that both males and females defended such territories in feeding flocks, usually on the margins. He noted the posture adopted by the territory owner when his territory was invaded, the owner running at the interloper in this attitude, fights with birds striking at one another with their wings and feet and displays with open wings. Lind (1957) described similar behaviour in autumn flocks in Denmark. Territories there were small, only two to three metres in diameter and mainly situated along a muddy shore. He observed that the crouching display constituted a threat when facing an opponent but turning and displaying the undertail coverts inhibited aggression and that birds often separated after a fight by both retreating in this posture.

Figure 7.2. Display in winter feeding territories; (i) two birds walking parallel along an invisible boundary in high upright pose. (ii) hunched or crouching run. (iii) fighting.

NOCTURNAL FEEDING BEHAVIOUR

Lapwings are often active at night. As a boy and a young man I was an avid wildfowler and learnt then how much Lapwings move about at nightfall in winter, often in ones and twos rather than as flocks going to roost. Many studies have observed a lunar rhythm in feeding behaviour in winter, with a decline in diurnal feeding activity around the full moon, implying nocturnal feeding. Feeding at night confers several advantages on Lapwings and food intake is considerably higher then (p.106). Such activity is necessary in most circumstances if the birds are to maintain themselves in winter. Nocturnal activity declines with frost (see below).

The lunar rhythm in winter feeding activity was obvious in my study in Sussex (Figure 7.3). The same pattern was repeated each winter but more birds fed during the day throughout the cycle in 1985/86, the coldest winter of the three, with much frost. Besides the obvious point that fewest birds fed at the full moon, the figure shows that some birds were nearly always either feeding or loafing by day at every point of the cycle. More birds also tended to feed by day at the quarters than at the new moon.

A very similar pattern was shown by counts I made in the winters of 2004/05 and 2005/06 of an isolated population in Meirionnydd, which roosted on a small tidal lagoon and fed on the surrounding grassland. I made counts on 38 days during November, December and January in the first winter and 42 days in the same period in the second. The birds were counted in the mornings up to 13:00 hours GMT. The pattern revealed by these counts is shown in Figure 7.4.

Although the same link with the lunar cycle was clearly shown by these counts, a significantly higher level of nocturnal feeding was suggested in 2004/05, which was a very mild winter, with long spells of settled weather. In 60% of the counts that winter all the birds were roosting, suggesting that they had fed during the previous night. All were feeding on only five days, three of which were exceptional in having frost and some light snow in the night. In 2005/06, by contrast, there were week-long spells of hard frost in each month, interspersed with mild, wet weather. In these colder conditions a higher level of diurnal feeding was very obvious.

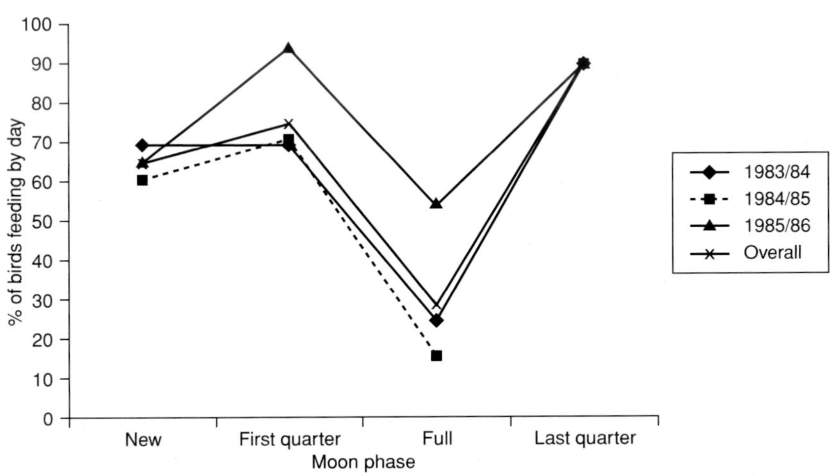

Figure 7.3. The relationship between diurnal feeding activity of Lapwings and moon phase on 873ha of mixed farmland in West Sussex, 1983/84–1985/86. Counts of feeding birds were done weekly from 10:00–13:00 hours from late October to early March. Counts plotted were those which fell within one day either side of each phase date. 31 out of the 44 counts made qualified but no count fell on the last quarter in 1984/85. Shrubb: previously unpublished data.

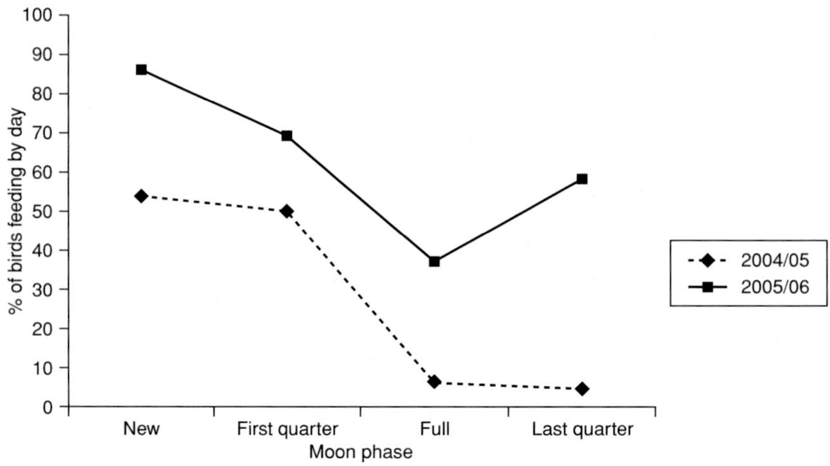

Figure 7.4. The relationship between diurnal feeding activity of Lapwings and moon phase on grassland in Meirionnydd. See text for methods. All counts within three days of each phase date were assigned to that phase.

January 2005 was also mild and frost free in northwest France and Trolliet (2005) found that, in these conditions, the plovers recorded in his census fed principally at night and sometimes entirely so. As found by Gillings (see below) distribution differed by day and by night and habitat use may also have done so.

The habit of nocturnal feeding has been long known (e.g. Yarrell 1845) and the lunar rhythm involved was apparently first recorded in 1882 by R. Payne-Gallwey (quoted by Milsom *et al.* 1990). Dementiev & Gladkov (1969) also commented that many of the invertebrate groups taken in steppe habitats are nocturnal. Only two studies, however, have specifically investigated nocturnal behaviour by Lapwings: Milsom *et al.* (1990) and Gillings (2003 and Gillings *et al.* 2005).

Gillings studied it over two winters on his arable site in Norfolk, using image intensifying equipment and infrared light, making some interesting and important discoveries. He found that Lapwings fed at night throughout the lunar cycle and that the percentage of birds doing so was linked not so much to moon phase as to moonlight, for numbers feeding did not differ statistically at the new and full moon but more Lapwings fed with increasing moonlight, along a trend from no moon to clouded moon to clear moon. There were seasonal variations, with more nocturnal activity in October and November than later, and nocturnal feeding activity declined with low ground temperatures, which partly, but not entirely, explained the seasonal pattern. Feeding also increased with time after sunset. Milsom *et al.* (1990) noted a rather different seasonal pattern in Hampshire, where evidence for the lunar rhythm was marginal in September and October; it was lacking in summer.

Gillings also found that nocturnal feeding flocks of Lapwings were rarely accompanied by Golden Plovers, with only 5% of nocturnal flocks comprising both species compared to 40% by day. Mixed-species flocks were also dominated by Lapwings but mixed diurnal flocks were equally likely to be dominated by either species. Nocturnal flocks were also much smaller than diurnal feeding flocks, although they were larger at the full moon than at the new.

Patterns of dispersion also differed. Feeding flocks were more widely spread at night, regularly using many fields never used by day. Only 27% of the fields checked were not used at night, compared with the 45% of fields which were not used by day. Fields used by day were also used, often more frequently, at night. Overall numbers in Gillings study area were also lower than expected at night and movements of up to 15km were observed between day- and night-feeding areas. Habitat use also tended to differ, with a stronger nocturnal selection for sugar beet and bean stubbles and lower use of cereal fields, where c.22% of feeding birds were found at night compared to 40–60% by day. As by day, grass was avoided for feeding at night.

Gillings *et al.* (2005) concluded that if these results were widely applicable, which is surely so given the frequency with which the link between feeding and moon phase has been noted, conservation proposals for site and habitat protection for wintering Lapwings (and Golden Plovers) cannot rest only on diurnal observations. These will describe neither their habitat nor range adequately. This is reinforced by the fact that Gillings (2003) found that diurnal feeding alone did not meet the daily energy needs of wintering birds. His was an arable site where earthworms, the most profitable prey, tend to be less numerous than in grassland. However, Barnard & Thompson (1985) formed the same conclusion on their pasture site, where they also noted nocturnal activity and where earthworms were the principal prey, and Fuller (1993) found that nocturnal feeding was regular in pasture in the Vale of Aylesbury.

DIET

Appendix 3 lists prey recorded in a number of European sources, largely derived from analyses of stomach contents or faeces. Such analyses involve some problems of interpretation as they may be biased towards prey such as beetles, which leave hard skeletal remains, and under-record soft-bodied prey such as larvae or slugs, which are digested more quickly and leave little or no trace. Some historical accounts suggest that items such as slugs, leather-jackets and wireworms may have formed a greater proportion of the diet in the pre-pesticide age. Field observations of feeding birds may also overstate the frequency of earthworms because it is easier to see one being taken than to be certain of a smaller item. On the other hand Lapwings often break earthworms in extracting them and eat pieces, which may then appear as different items. Another source of potential bias is that many published diet lists give little exact indication of season, although a high proportion of those examined appear to be for the breeding season. This is also a source of bias as summer diets tend to include a greater proportion of surface-dwelling invertebrates as earthworms, particularly, become increasing inaccessible as the soil dries out in summer and autumn.

Nevertheless Appendix 3 shows that a very wide range of invertebrates is taken. In the samples tabulated beetles (including weevils), earthworms, slugs and snails and Diptera were the most frequently reported items, with Lepidoptera larvae secondarily. A surprising number of samples included vegetable matter, particularly seeds. How far these are deliberately eaten or accidentally ingested is unclear. Fruit is certainly taken in some habitats, e.g. cranberries in peatbogs in Estonia (A. Kumari in Hale 1980).

The diet has many features in common with that of other ground-feeding birds, particularly corvids, gulls feeding in farmland, thrushes such as Fieldfares and, to some extent, Starlings. Whether interspecific competition for food exists between these species is unclear. My Sussex records suggested that Rooks fed in the same fields as Lapwings rather infrequently but gulls (which parasitise Lapwings), other waders, thrushes and starlings did so frequently.

Although the invertebrate groups upon which Lapwings feed are abundant and widely distributed, various seasonal, habitat related, geographical and age differences in diet seem clear. Seasonal variations largely reflect the annual cycles of prey organisms and seasonal feeding conditions – the increasing dryness of the soil in summer and so forth. Earthworms appear to be most important in winter and spring/early summer, becoming increasingly inaccessible as soils dry out in summer and early autumn. They are particularly significant to females during the pre-laying period, when they accumulate the resources for a clutch (Högstedt 1974, Galbraith 1989b, Baines 1990). Their overall importance to Lapwings was underlined by experience in The Netherlands in the very dry summer of 1959. Drought lasted from mid-May to mid-October and birds were unable to feed on earthworms. As a consequence birds collected were in very poor condition, with body weights up to 48% lower than normal. Birds did not flock for feeding and resorted to foraging in unusual sites such as roadsides and lawns. Increasing numbers of dead birds were found (Voous 1962). M. F. Peers (pers. comm) watched migrant Lapwings in the Pyrenean foothills in March 2005 which were feeding along roadsides because of severe frost and very dry conditions.

Baines (1990) found a marked seasonal change of diet in northern England, with a sharp decline in the numbers of earthworms, beetle larvae and Diptera taken after 30 April and a marked increase in adult beetles. E. Blezard (in Cramp & Simmons 1982) noted similar changes in Cumbria (Appendix 3). Similarly in coastal grazing marsh in Kent, adults took progressively fewer earthworms and tipulid larvae during April to June, with increases in Lepidoptera larvae, chironomid and stratiomyid larvae and Coleoptera. Chicks there fed extensively on Diptera larvae but earthworms were important after a wet spring in 1994 (Appendix 3). As with earthworms, leather-jackets (tipulid larvae) are chiefly winter and spring prey (e.g. Galbraith 1989b) and noctuid moth larvae are perhaps mainly taken in autumn, which again presumably reflects the prey organism's life cycle; species such as the

Turnip Moth would only be readily available during the autumn before crops were lifted or grazed off.

Geographical variations are not particularly obvious, although species of groups such as Coleoptera must replace one another across the range. The data in Appendix 3 suggest, however, that surface invertebrates form a much greater proportion of the diet in the east, for example in the former Soviet Union. As Lapwings are summer visitors to that region, however, this may be a seasonal rather than a strictly geographical difference. One marked difference, however, is that Acrididae (grasshoppers) appear only to be noted as prey in steppe habitats in eastern Europe and Russia. They have rarely been recorded in the diet in improved farmland or rough grazing in Britain or elsewhere in western Europe.

Habitat-based variations in Lapwing diets are more marked. For example in winter in Britain earthworms are more frequently taken in pasture and may be virtually the only prey taken there. In arable land, wintering birds always take a much wider range of prey, with many surface invertebrates, although earthworms still often provide the greatest proportion of the diet by weight. Leather-jackets are perhaps particularly taken in leys. Lapwings exploiting estuarine habitats feed on a different range of animals. These are often perhaps equivalents, marine worms such as Nereidae (ragworms) instead of Lumbricidae, marine crustacea and so on.

Habitat-related differences in diet may be particularly marked in chicks. Galbraith (1989a) found marked differences between the diets of chicks in arable land (which included leys) and rough grazing, with significantly more leather-jackets and fewer coleopterans taken in arable land. Similarly Matter (1982) found significantly different proportions of Coleoptera, insect larvae and earthworms in the diets of chicks in arable land in Switzerland and mixed farmland with grassland in north Germany (Appendix 3). In Sweden chicks in shore (saltmarsh) pastures fed extensively on the marine worms Nereidae and took few Diptera, which were important to chicks on fresh pasture. Both took similar numbers of coleopterans but of different species (Johansson & Blomqvist 1996).

FOOD INTAKE AND FACTORS AFFECTING IT

One difficulty inherent in calculations of energy intake in Lapwings is that earthworms, although the most important source of energy for wintering birds, form only part of a diet which, in arable land particularly, may comprise many more small surface invertebrates and, in some habitats, more significant items such as leather-jackets or slugs, which are difficult to quantify adequately. Most studies have assumed that earthworms are overwhelmingly the most important prey because of their size disparity with other invertebrates but Gillings (2003) pointed out that the general assumption that prey are selected to maximise energy intake may not always be true. Factors such as the need for protein or other essential nutrients may lead to different choices.

Gillings (2003) estimated that Lapwings need 500–550kJ of energy per day, which required a daily intake of 27g dry weight of food for subsistence. Daily intake was, in fact, variable and the rate depended very largely on the number of earthworms taken and their size. Calculations based on the actual frequency of worm sizes taken gave a predicted daily intake requirement of 278 earthworms to meet subsistence needs. Gillings noted that predicted intake levels, based on actual activity levels ranged from 0.80–2.40mg of Ash-free Dry mass (AFDM) per second but actual levels ranged from 0.30–0.73mg per second. Actual intake thus fell below the rate required for subsistence. Furthermore the predicted levels calculated assumed that Lapwings fed throughout the daylight hours available, which is often not so. In full moon periods diurnal intake provided no more than 30–50% of daily needs.

Low levels of diurnal intake may be typical of Lapwings. For example, Gillings calculated levels of 0.07–0.52mg AFDM per second from Barnard & Thompson's results on their pasture site (Barnard & Thompson 1985), where diet was primarily earthworms. The shortfall has to be made up by nocturnal feeding. Darkness does not appear to be a major problem for foraging Lapwings, which have good night vision (p.94) and may be able to locate prey by hearing. Nocturnal feeding confers several advantages on Lapwings. In particular it frees them from the piratical attentions of gulls. Also important prey, particularly earthworms and especially large earthworms such as *Lumbricus terrestris*, are much more readily available. Thus nocturnal food intake rates are much higher and may be twice those of diurnal feeding.

Intake may also be constrained by the need for digestion. Gillings (2003) quoted work showing that this was the case with the intake of bulky prey items by Oystercatchers and Whimbrels and suggested this may also be the case with plovers taking earthworms. I have certainly noted this with Common Buzzards feeding on earthworms, a regular habit in winter in Wales, which exhibit a very marked pattern of bouts of feeding interspersed with substantial periods of rest, presumably for digestion.

Farm management also has a significant impact on the populations of those soil invertebrates which are of potential importance to Lapwings and other birds feeding on them. The main factors involved seem to be the frequency of cultivations and fertiliser inputs, particularly of nitrogen and farmyard manure. In grassland, grazing intensity, stocking patterns and types of grazing animal may also be important, as may variations between grazing and mowing. In arable land the return of organic matter to the soil may also be an important feature.

In both habitat categories cultivations, particularly ploughing, have a traumatic impact on invertebrate populations. Edwards (1984) noted good evidence that it can decrease populations of soil invertebrates by up to 50% and they may take several years to recover. Ploughing and reseeding old grassland invariably replaces mixed species swards with single species ones, which also reduces variety in invertebrate populations. Reseeding also alters sward density and may thus reduce the value of grassland areas as cold weather refuges, particularly if frequently done, because soil insulation is poorer.

Heavy grazing and high applications of inorganic fertiliser, particularly nitrogen, also reduce plant species diversity in grassland and the populations of many invertebrates are inversely related to nitrogen application rates. However, except at very high rates, earthworms benefit from grassland improvement and nitrogen applications and so do beetles other than carabids. Applications of farmyard manure maintain invertebrate populations and encourage greater activity at the surface but heavy applications of slurry, the form which farmyard manure now often takes, can be highly toxic to earthworms.

In arable land the return of organic material to the soil is an important additional factor. Although straw burning, which Edwards (1984) showed significantly reduced soil invertebrates, is now banned in Britain there has probably been a decline in this return today. With rotation farming in the past, although most straw was removed with the sheaf, it was returned in dung from livestock and with ploughing-under leys at intervals. Leys themselves also provided a one- to three-year break from cultivation, allowing invertebrate numbers to recover. The use of herbicides not only reduces the volume of organic material returned, but also the habitat and therefore abundance of many arthropods. High levels of nitrogen usage, however, seem to benefit invertebrates in arable land, probably because they lead to an increase of organic material returned to the soil. (Summary based on Edwards 1984, Baines 1990, Beintema *et al.* 1991, Tucker 1992, Sotherton & Self 2000, Wakeham-Dawson & Smith 2000, Vickery *et al.* 2001, Buckingham *et al.* 2006).

Overall the intensification of management in grassland and arable land, based on reseeding, high fertiliser rates, herbicides and changes in rotations, have tended towards lower invertebrate populations. The impact of increased fertiliser rates is mixed and may be most marked in summer because their effect is greatest on surface invertebrates; soil invertebrates, which are more important to birds at other seasons, tend to benefit from increased fertiliser application rates (e.g. Atkinson *et al.* 2005). Nevertheless, whilst evidence that these changes affect wintering Lapwing populations is lacking, it seems very likely that at least some of the variability in field use by feeding birds identified in various studies derives from such variations in management. Lapwings are pioneers and their feeding behaviour is strongly opportunistic (p.86, Table 6.4) which, together with the consistent tracking of the rotation in my study, suggests that they are adept at teasing these variations out and exploiting them.

FOOD PIRACY BY GULLS (KLEPTOPARASITISM)

Gulls, mainly Black-headed Gulls, are important kleptoparasites of feeding Lapwings (and Golden Plovers). Such piracy has several marked and constant features. In Britain it occurs far more frequently in areas where plovers feed in pasture because the gulls steal earthworms. Plovers, particularly Lapwings, take time to extract and manipulate earthworms, becoming vulnerable to attack. When they feed on smaller items or surface invertebrates gulls rarely bother them, as these items are picked up and swallowed too quickly to make stealing profitable. In arable habitats, therefore, piracy is relatively infrequent and the presence of gulls in a Lapwing flock is a good guide as to whether earthworms are being taken regularly. If they are not and waiting is unprofitable, the gulls soon drift away.

Where gulls are feeding by stealing from plovers they tend to select sites where plover density is high and they space themselves through the plover flock at distances of 5–20m between each bird, maintaining that spacing by threat and aggression (Plate 25). Effectively this results in a series of mobile territories scattered through the plover flock. Enough gulls may be present for this system to cover the entire plover flock. Gulls undertaking long chases may then have difficulty in returning. Kallander (1977) found that the number of Lapwings to gulls could be as few as 2:1 but was usually much more. In Wales I have found that ratios of less than 4:1 will cause Lapwings to move elsewhere. Feeding efficiency for gulls depends on the plovers taking a particular size range of earthworms. Small earthworms, like surface invertebrates, are handled too quickly to make theft profitable. Searching for targets is often done from a slightly raised stance, such as a molehill. When a plover finds a suitable prey item the gull flies at it, whereupon the plover either drops the item or attempts to fly off with it and evade attack, prompting a chase which usually results in the worm being dropped and taken by a gull. Attacks may be launched over distances of 20m or more. Gulls are more successful against Lapwings than against Golden Plovers because the latter are faster and more agile on the wing.

The numbers of gulls that can feed by piracy must be fairly insignificant in terms of their total population and it is also limited by their own behaviour. Nevertheless Kallander found within his Swedish study area that 44% of the Black-headed Gulls present were feeding by piracy in the spring and 27% in autumn. He found that, in spring, piracy was only slightly more frequent on pasture than in habitats such as plough and stubble and it was most frequent in oil-seed rape crops. Although such piracy is mainly a winter activity in Britain, Kallander noted it particularly during spring Lapwing passage in southern Sweden.

Black-headed Gulls can support themselves by such theft. Kallander recorded that a mean of 0.272 earthworms per gull per minute were taken, the equivalent of 163 earthworms in a ten-hour day or 200kcal (840 kJ) per day, well above the energy intake gulls required, even allowing for the expenditure incurred in chases. Such kleptoparasitism is a costly process for Lapwings. They, in any case, fail to meet their daily intake requirement by diurnal feeding alone but, as Kallender's calculation above suggests, that deficit is greatly increased by the piratical operations of gulls.

Plovers do have defences against such piracy. Lapwings (and Golden Plovers) change their intake of different earthworm sizes in response, concentrating more on smaller sizes which can be handled quickly, although this also reduces the profitability of their feeding. They also move away from watching gulls, so the presence of gulls reduces the density of feeding plovers.They turn their backs to gulls, making it more difficult for gulls to observe their behaviour. They also handle large earthworms only when a long way from gulls. Indeed, Lapwings often discard larger earthworms anyway, presumably to avoid being chivvied, which involves energy expenditure. Ultimately, as noted above, pressure from gulls will cause plovers to move to new feeding sites and the cycle starts again. (Summary largely based on accounts by Kallander 1977 and Barnard & Thompson 1985).

Other species may attempt to steal food from Lapwings. In Wales I have noted that Carrion Crows do so frequently but they are far less successful than gulls. Lapwings' reactions to them are much closer to their reaction to raptors and a crow flying through the flock hoping to steal prey will cause the entire flock to take off. Gillings (2003) also recorded piracy by Starlings, which I have never observed.

CHAPTER EIGHT

The breeding season: arrival and territory

As the time for departure from their wintering grounds approaches, Lapwing flocks often become restless, frequent bickering and sudden flights round the flock introducing a ripple of disquiet. Birds chase or run at each other in the threat attitude used in defending winter territories (p.100). When chased, birds jump up into the air for short flights, flashing their white wing-linings, and sometimes they do a sequence of short flight chases or hops in this way. Individuals will run through the flock with their wings raised, showing the wing-linings. They stand very straight facing each other displaying the breast band and occasionally opening their wings, again showing the white underwings. Birds regularly make rapid erratic flights round the flock, which often develop into incomplete song flights. These are clearly embryo display flights and incomplete songs may be given. All these movements and actions are part of the territorial and courtship behaviour to come.

The pattern of spring movements and arrival is discussed in Chapter 13 but spring arrival on the breeding grounds is early, for example from mid-February in Britain with little variation from south to north. Dementiev & Gladkov (1969) noted that the first birds arrived in the Soviet Union as soon as small thawed patches appeared in the fields and sometimes before. Sudden cold snaps, which are not uncommon on the fringes of winter, sometimes lead to disasters (Chapter 13) and often reverse the pattern of arrivals. Dementiev & Gladkov, for example, recorded early arriving birds which encountered severe frost vanishing for a further three weeks, behaviour which Rogacheva (1992) noted as annual in central Siberia. Galbraith (1989b) noted temporary reductions in his breeding populations with spring cold snaps in central Scotland. Snow and severe frost in the first five days of March 1965 triggered a typical and large westerly cold weather movement of Lapwings along the Sussex coast, sharply reversing the spring arrival. This was by far the latest date I ever recorded for such an event.

Both Spencer (1953) and Nethersole-Thompson & Nethersole-Thompson (1986) noted how unobtrusive the spring arrival on breeding grounds is. On first arriving flocks and parties tend to stay together, feeding on the fields, often those outside the eventual territorial systems. On his sites Baines (1990) found that 75% of these flocks comprised females, which used certain high-quality neutral feeding fields to acquire the resources needed for their clutch while undisturbed by territorial mayhem. Males first visit their territories at dusk and dawn and often display at night, particularly on moonlit nights. Gradually they spend more time there until the system of territories is formed.

Males often arrive earlier on territories than females. For example the Nethersole-Thompsons noted that some males appeared up to a week before females in Speyside and I have noted that males may appear on territory up to two weeks before females in Wales. In southwest Norway males were up to ten days earlier than females (Grønstøl 1996). Högstedt (1974) noted that pairs in his study area in southern Sweden were mated before or immediately after occupying their territories but in other territories nearby females appeared on their future breeding sites up to three weeks later than males. Such patterns of female settlement seem typical of my Welsh site, as laying dates suggest (p.137) and an isolated 'pair' there in 2006 provided an excellent example of what happens. This pair had mated and the female was incubating by 2 April but she appeared to have two males (one of them colour ringed) attached to her, although I only saw her mate with the unmarked male. I found a second female on the site on 14 April, which mated with the marked male and started incubation on the 20th. On 16 April a third female had appeared, which eventually mated with the unmarked male and started incubation on the 24th. He now had two females, one close to hatching a clutch and one just starting incubation, but the first female lost her nest just before hatching. She was incubating a fresh clutch on 7 May. As also noted by Baines (1990) in northern England, in the early stages of territorial establishment before settling, these females remained with a flock feeding on sheep pastures nearby.

Once settled on territories birds will sit out sudden spring cold spells. Baxter & Rintoul (1953) observed that 'We have often noticed how loath they are to leave their breeding grounds once they have returned, even though late storms may cover the fields in which they are going to nest with a considerable depth of snow'. Nelson (1907) recorded that 'in a severe storm in May 1891, near Harrogate, the Lapwings flew high above the snow clouds; when the fall ceased they returned to their breeding quarters, but were unable, on the snow-covered ground, to find their nests'. Such events are recorded in the BTO nest record cards into the 1980s. As noted above, further north and east severe weather in spring can cause significant losses in local breeding populations.

TERRITORIAL SYSTEMS

Lapwings are strongly territorial birds. Territory sizes seem surprisingly small and in Britain various sources have given mean sizes ranging from 0.3–0.9ha (Howard 1920, Spencer 1953, Nethersole-Thompson in Cramp & Simmons 1982, Redfern 1982). In Sweden Berg (1993) found that the mean size of 20 randomly selected territories in mixed farmland was 1.6 ± 0.9ha, with polygynous males holding a mean of 2.0 ± 0.8ha and monogamous males a mean of 1.1 ± 0.8ha. Males in colonies held smaller territories than those nesting solitarily. Byrkejedal *et al.* (1997) indicated smaller territory sizes of 0.1–0.75ha on an area of pastoral farmland in Norway.

Territorial systems are variable. Probably the preferred form of nesting dispersion is in loose social groups, which can become semi-colonies in areas of high population density. Within such groups males maintain clearly defined individual territories, which provide discrete secure sites in which courtship, pair formation, nesting and feeding can take place comparatively undisturbed. Territory may also contribute to population regulation by pushing some settlers into poorer habitats where breeding success is low. Selection of nesting territories tends to be based largely on the habitat features which aid nest security – cryptic backgrounds for concealment and open fields of view to facilitate active nest defence. Although it may underlie variations in nesting density (Table 8.1 below), food supply is less important in territory selection. Where the supply is good the territory provides all the food needed by adults although, after hatching, broods may disperse beyond their natal territories. Where food supply is sparser, neutral feeding grounds away from the nesting territory are used by adults and young. Although the Nethersole-Thompsons recorded occasional movements of up to 7km to feed by off-duty birds, usually females, such neutral feeding sites usually involve movements of little more than *c.* 300m and nesting territories may still be grouped or semi-colonial. Such duality in nesting territories, with one habitat for the nest and another for feeding and rearing the chicks is widely recorded in Lapwings, particularly in arable land which may provide attractive nest sites and a good early food supply for females but poor resources for chicks, which are moved to pasture (Klomp 1954, Cramp &

Simmons 1982, Nethersole-Thompson & Nethersole-Thompson 1986, Galbraith 1988c, Baines 1990, Berg 1993, Byrkjedal *et al.* 1997, pers. obs.). As Blomqvist & Johansson (1995) pointed out, selecting a nest site and rearing the young thus involves a trade-off between the benefits of nesting close to good feeding sites for the adults and the costs of moving chicks longer distances to suitable feeding areas.

Selection of nesting territories within a given area may change with habitat and management of the site. Thus on my family farm in the 1950s and early 1960s a population of 20–30 pairs was mainly distributed in groups of 2–12 pairs. One group of up to seven pairs was located in a large area of old grassland and they maintained themselves within their territories. Otherwise pairs were mainly concentrated on arable fields, always adjacent to pastures which were used as neutral feeding areas and for rearing chicks. This limited choice to seven arable fields and only three other arable fields were ever used on the farm, once each, in 30 years. These patterns of territory and dispersion changed as agricultural productivity on the farm was increased by drainage, higher fertiliser inputs and a switch to autumn cultivation from the mid-1960s. Before that the traditional fields were generally used whatever the crop, as these fields had areas of rather poor soils and most crops in them provided the necessary thin areas of vegetation for nest sites (Chapter 10). Nevertheless bare fallows for roots and clover crops, which were really under-sown stubbles when Lapwings settled, were the most favoured; autumn cereals were used rather more often than spring. As the management of autumn cereals changed (p.55) the Lapwings increasingly switched to nesting on pasture and spring cereals, other spring-tilled habitats no longer being available, and, as the management of spring cereals intensified, eventually they increasingly moved to pasture (Figure 8.1). On pasture they maintained themselves within their territories as in the 1950s, although chicks were often shifted for rearing, particularly to the vicinity of pools and splashes.

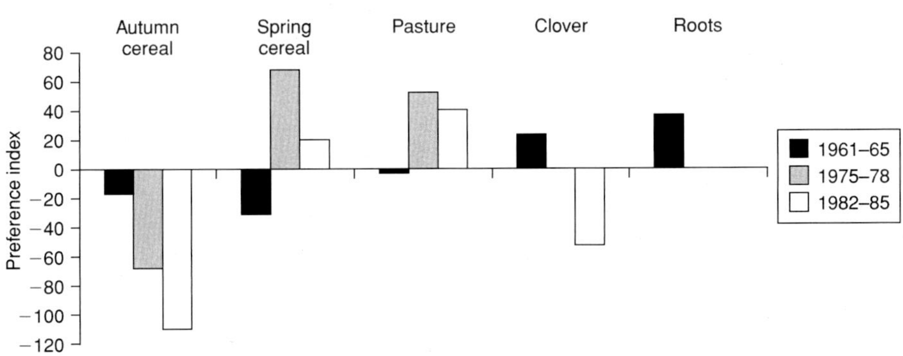

Figure 8.1. Habitat selection by nesting Lapwings in three periods with different rotations at Oakhurst Farm, West Sussex. Root fields were bare tillage when Lapwings were nesting. Root crops were discontinued after 1965. Field beans and clover were grown in the 1970s but were not used by Lapwings. Preference index as in Table 4.2. Points above zero indicate that the habitat was selected, points below zero that it tended to be avoided.

NESTING DISPERSION

There seems to be little historic information on the general patterns of dispersion in breeding Lapwing populations. In Britain, one can guess that groups and colonies were once the typical form of dispersion in most lowland populations breeding in the core habitats of the Waste (p.44). Although hedges had long existed in British farmland they were not a feature of these habitats, which were most often managed under rights in common, which barred enclosure and fencing. They were typically very open and maintained by grazing. Their progressive enclosure in the late 18th and early 19th centuries introduced a lattice of field boundaries, largely hedges with trees. Enclosure also introduced an extensive infrastructure of roads, farmhouses, buildings and so on. Pell (1887) quoted the example of one 1,648 acre (760ha) common, the enclosure of which involved the installation of 5km of parish roads, 1.6km of occupation roads, 57.5km of quickthorn field hedges and 8.3km of boundary and road hedges, besides farmhouses and buildings. This gives a clear idea of the sheer scale of landscape change wrought by enclosure.

Whilst their impact would not have been immediate, the presence of trees, hedges, farmsteads and roads limits the utilisation of fields by Lapwings (van der Zande *et al.* 1980, Elliot 1982, Reijen *et al.* 1996, Sheldon *et al.* 2004). Perhaps inevitably, therefore, dispersion and field size have become linked. Figure 8.2, based on data drawn from the 1987 BTO Lapwing survey, shows a strong link between field size and the numbers of pairs of Lapwings present in occupied fields.

The general patterns of dispersion in these surveys is summarised in Figure 8.3. In 1987, 788 of the 1,224 fields in the sample held single pairs, 280 held

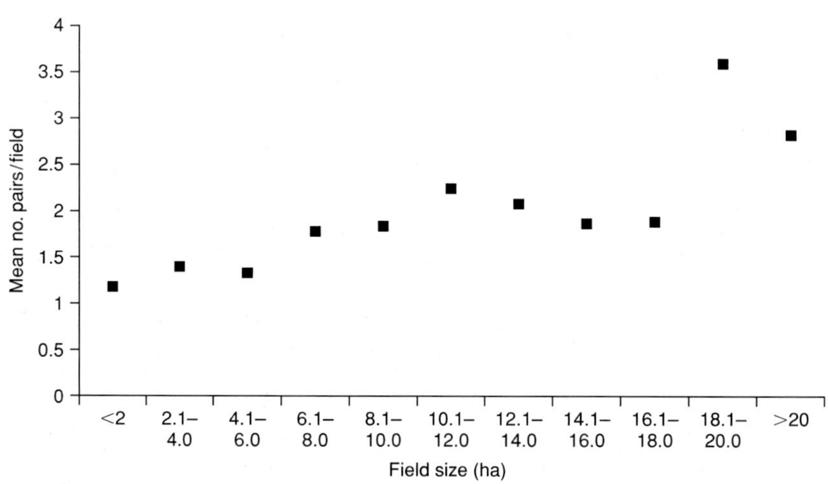

Figure 8.2. The mean number of pairs per occupied field in fields of different sizes in England and Wales in 1987. Data from 1,224 fields for which size and the number of pairs was recorded (Shrubb & Lack 1991).

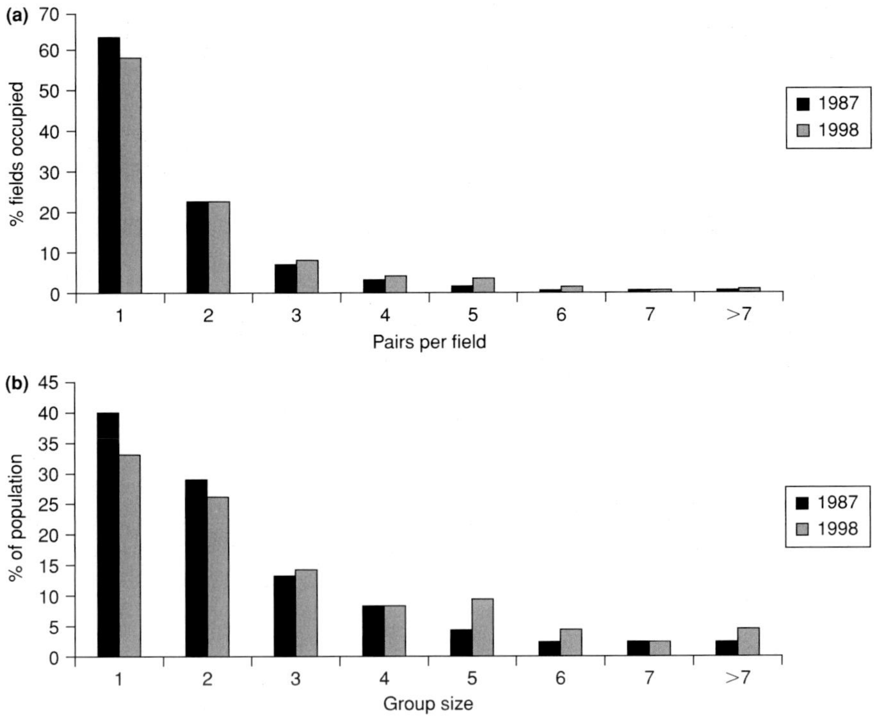

Figure 8.3. Dispersion of nesting Lapwings in a sample of 1,224 occupied farm fields in England and Wales in 1987 and 799 in 1998: a) the % of fields containing different numbers of pairs; b) % of the population occurring in groups of different size. Data from Shrubb & Lack 1991 and maps from the 1998 BTO Lapwing survey.

two pairs and 156 three or more. If representative, that indicated that c.40% of the population was breeding as single pairs, *c.* 29% in groups of two and *c.* 31% in groups of three or more. In 1998, when the population had halved, 474 out of 799 fields held single pairs, 185 two pairs and 140 three or more, giving 33% of the population occurring as single pairs, 26% in groups of two and 41% in groups of three or more. Thus declining numbers were associated with a tendency to occur in larger groups. The figure should be regarded with some caution, however, as group size is taken as the number of pairs nesting in each individual field. But groups may involve birds nesting in several neighbouring fields, particularly where field boundaries are very open. This was particularly so in some, mainly pastoral, areas with rather small fields, where single pairs occupied several neighbouring fields, especially in the north in 1998. Effectively this is another form of loose association and the proportion of genuinely solitary pairs within the population is probably less than 30–40%, perhaps considerably so. It seems likely, in fact, that such pairs are often unsuccessful and unlikely to persist, which perhaps explains the tendency for group nesting to increase by

1998. Small fields were also more often occupied in areas of high density in England at least.

In 1987 the only obvious link between dispersion and general habitat categories was that there were more groups of two or more pairs in fields under spring tillage and more single pairs in permanent grass fields than expected (Shrubb & Lack 1991). As mean field size was about one third larger in tillage, this was perhaps a factor of field size rather than habitat. Although regional samples were often rather small, their analysis suggested that patterns of dispersion did not vary significantly across regions in either grassland or tillage. Field size was not examined in the 1998 survey.

Studies elsewhere have revealed very similar patterns of dispersion. For example in Aberdeenshire Elliot (1985a) found about 20% of his pairs nesting singly and the rest in loose groups of 2–5 pairs, occasionally up to ten. Two studies in German farmland found 29% and 54% of pairs nesting solitarily, 33% and 49% distributed in groups of 2–4 and 14% and 22% in groups of 6–10 (H. Prill and K. Kirchhoff in Cramp & Simmons 1982). In central Sweden Berg *et al.* (1992) found that 7.8% of 870 nests were solitary, 35.1% were in colonies of 2–5 nests, 29.8% in colonies of 6–10 and 27.2% in colonies of >10, with a maximum of 28 pairs.

Thus a mixture of solitary pairs and group nesting appears to be the normal pattern of nesting dispersion in Lapwings. It may partly reflect the nature of the main nest predators of the species and its anti-predator strategies. Groups increase the effectiveness of nest defence against predators such as corvids but not against more dangerous predators such as foxes. Here passive defence through spacing cryptic nests reduces predation risk (Chapter 11).

Within groups or colonies nest spacings vary considerably. Berg *et al.* (1992) defined a colony as all nests within 200m of any other and close neighbours as those within 100m of a nest. Seymour *et al.* (2003) recorded nearest neighbour distances for 116 nests, which averaged 45m and ranged from 12–153m. Within colonies they found up to nine neighbouring nests within 100m radius of a nest, with a mean of 3.7. Kooiker (1984) noted the mean distance between 128 nests at Osnabruck Germany as 55.6m. Distances of ten metres down to two metres have been noted but these seem most likely to refer to polygynous matings. Thus in Wales in 2005, the nests of three females mated to one male were spaced along a line nine and six metres apart. All three were lost to flooding and the repeat nests were 50m, 30m and 10m from the original sites, with the first two 40m apart. It was interesting to watch the male apparently enticing his females to shift their nests into a drier part of the pasture through his scraping displays. Smith (1917) recorded two nests, each with two eggs, which were actually touching.

BREEDING DENSITIES

Breeding densities vary widely. Table 8.1 summarises a range of densities drawn from the literature. It shows considerable differences with habitat but the majority

1. A superb shot of a breeding adult male (Tomi Muukkonen).

2. A breeding adult female. Note the admixture of white in the black of face and chest band and much greener back (Aurélian Audevard).

3. A fine close up of an adult male, not quite in full breeding plumage (Rick van der Weijde).

4. Yellow Wattled Lapwing (Dave Behrens).

5. Red-wattled Lapwing (Nikhil Devasar).

6. Sociable Lapwing in summer plumage (Marten van Dijl).

7. Grey-headed Lapwing in flight (Sumit Sen).

4–7. Some lapwing species; long legs, facial adornments and bold plumage patterns are characteristic of the group.

8. An adult Lapwing in winter plumage (R.J. Wilmshurst).

9. Displaying male Lapwing. An early arrival with much snow still lying (Jari Peltomäki).

10. Two adult male Lapwings in territorial dispute, showing the characteristic wing shape with broad, rounded primaries (Peter Hadfield).

11. Female Lapwing at nest with newly hatched chicks (R.J. Wilmshurst).

12. Lapwing nest with a somewhat unusual clutch of five stone-coloured eggs (R.J. Wilmshurst).

13. A typical clutch of four eggs (Rick van der Weijde).

14a. Chick less than a week old. Note the size of the chick's leg and foot (Barry Yates).

14b. A fast-growing chick under a fortnight old (Mark Plomp).

15. A chick close to fledging, showing well the characteristic freckling of pale feather edgings on the upperparts in juveniles (Alice Stevenson).

16. The herringbone pattern of newly laid field drains (R.J. Fuller).

17. Arable nesting habitats. Peas in June, still suitable for nesting (M. Shrubb).

18. Arable nesting habitats. Maize in June (M. Shrubb).

19. Arable nesting habitats. Set-aside fallow holding 2 pairs in June (M. Shrubb).

20. Grassland habitats. Old wet grassland in Meirionnydd just before pairs settled (M. Shrubb).

21. High stocking densities in the same county. Here there are 250 ewes and lambs and 16 cows and calves on *c.*9ha, a stocking density of 30 animals/ha but a stocking rate of only 3.5 livestock units/ha (M. Shrubb).

22. Ridge and furrow (R.J. Fuller).

23. Tightly grazed sheep pasture in winter (R.J. Fuller).

24. Lapwings feeding by hedges and buildings (R.J. Fuller).

25. Characteristic distribution of gulls in a feeding plover flock; grass cattle grazed (R.J. Fuller).

26. Winter cereals in January sown at traditional spacings and timing (R.J. Fuller).

27. Winter cereals in January sown at narrow spacings and early (M. Shrubb).

28. Lapwing flock in flight (Aurélian Audevard).

29. An adult male bathing, showing the tail pattern well (John Robinson).

Table 8.1. Some breeding densities of Lapwings in Europe.

Country	*Site and year*	*Area (ha.)*	*Habitat*	*Density: pairs/km²*	*Reference*
Scotland	Dorback Moor 1942	260	moor, rough pasture, farmland	10.4	Nethersole-Thompson & Nethersole-Thompson 1986
	Orkney 1951	260	farmland	76.9	Spencer 1953
	Orkney 1974	1,098	moorland	1.83	Lea & Bourne 1975
	Outer Hebrides 1983 & 1984	18,600	machair	20–27	Fuller *et al.* 1986
	Carse of Stirling 1984/1985	856	mixed farmland	18.9/15.4	Galbraith 1989b
	Carse of Stirling 1984/1985	800	rough grazing	15.9/15.9	Galbraith 1989b
England	Oxford 1939	400	farmland	1.6	Nicholson 1938/39
	Cheshire 1930	400	farmland	4.5	Nicholson 1938/39
	Surrey 1930s	90	farmland	7.8	Lister 1939
	Sussex 1984	6,745	tillage	1.46	Shrubb 1985
	Sussex 1984	6,408	grassland	2.83	Shrubb 1985
	N. Kent Marshes 1990s	6,500	grazing marsh	15.23	Milsom *et al.* 2002
	Teesdale 1985–1995	412	pastoral & marginal farmland	80–88	Parish *et al.* 1997a
	Eden Valley 1990–92	465	pastoral & mixed farmland	47 declining to 24	Thompson *et al.* 1994
England & Wales	General 1956–65	5,483	farmland	6.4	See footnote
Germany	Bayern		river valley	0.9	Glutz von Blotzheim *et al.* 1975
	Schleswig-Holstein		river marsh	20	Glutz von Blotzheim *et al.* 1975
	Magdeburg		flooded/ not flooded land	0.49/0.91	A. Ulrich in Cramp & Simmons 1982
Belgium	Antwerpen-Linkeroever 1983–86		Fallow/cultivated land	62/162	Van Impe 1988
The Netherlands	overall		grassland	25–40	Hagemeijier & Blair 1997
Soviet Union	Vladimir region		fields, meadows	10–13	Rubinshtein 1968 in Cramp & Simmons 1982
	Central Siberia 1980		hay fields	1–9 birds/km²	Rogacheva 1992

Table 8.1. *Continued.*

Country	*Site and year*	*Area (ha.)*	*Habitat*	*Density: pairs/km²*	*Reference*
Sweden		100	shore meadows	50	Larsson 1976
	Vastmanland 1988	5,900	mixed farmland	4.3	Berg 1993
	Odsmals kile 1987		Coastal farmland	37	Johansson & Blomqvist 1996
	Valida Sando 1987–1990		Coastal farmland	54	Johansson & Blomqvist 1996
	Torkelstorp 1987–1990		Coastal farmland	35	Johansson & Blomqvist 1996
Denmark	Aggersund 1977–79	20.0–27.5	saltmarsh	80–110	Ettrup & Bak 1985
	Logstor 1978–79	17.0–37.5	wet saltmarsh	76–83	Ettrup & Bak 1985
	Alro 1977–79	40	saltmarsh and arable land	38–43	Ettrup & Bak 1985
	Kolindsund 1978–79	400	arable land	4–6	Ettrup & Bak 1985
	Tipperne NR 1985–1992	700	brackish meadows	32–39	Thorup 1998
Finland	overall 1984		farmland	5–13	Piiroinen *et al.* 1985
	Pyhäjärui 1966–1968	1,373	peat bog raised bog	2.3–2.9 6.9	Hakala 1971 Hakala 1971

The records for England & Wales in 1956–65 were drawn from the original papers from the survey by Lister (1964); see Shrubb & Lack 1991. See also Chapter 4 for densities in grassland nature reserves in Britain and Figure 2.2 for overall densities in farmland areas across Europe.

of the highest densities comprise areas of wet or damp grassland. One problem with such density figures is that they tend to be drawn from study areas selected for reasons other than examining population density, rather than from randomly selected samples. So they cannot really be used for historical comparisons. In general, however, there is a predictable tendency for recorded densities to be lower on larger sites: larger areas are likely to include more unsuitable habitat. In England and Wales as a whole, the population density in farmland recorded by the 1987 Lapwing Survey was 1.09 pairs/km², which had declined to 0.6 pairs/km² in 1998. An approximately contemporary survey of lowland Scotland in 1992/93 found an overall density of 2.2 pairs/km² (O'Brien 1996).

Some much higher breeding densities have been recorded. For example Aplin (1889) recorded 50 nests in one 30 acre (12ha) field in Oxfordshire and Smith (1887) recorded 40 nests being destroyed in the harrowing of one field in Wiltshire. Pyefinch & Golborn (2001) recorded 38 pairs in one arable field in Lancashire in 1992 and also a group of 12 pairs on a 2ha island in Belmont Reservoir there. Ryves (1948) found ten pairs on a 1.5ha field in Cornwall, Nethersole-Thompson &

Nethersole-Thompson (1986) reported a record of 15 nests on half of a 5.5ha field in Lincolnshire in 1930 and Shrubb (1979) recorded 14 pairs nesting in 1.3ha of kale in Sussex. The nest records of the BTO include one of nine nests on a 1.6ha patch of rough grass in the corner of a barley field in Kent. In Norway Byrkjedal *et al.* (1997) studied a population of 22 males and 36 females nesting on 6ha. Such figures should perhaps be regarded as measurements of colony size rather than breeding density and the British examples quoted above provide a good example of the flexibility of territorial behaviour, enabling groups of Lapwings to exploit very small areas of good nesting habitat.

MATING SYSTEMS

The Lapwing has traditionally been regarded as monogamous and single brooded, with polygamy occurring rarely (Cramp & Simmons 1982) but with a strong tendency to promiscuity. Recently, however, studies using a combination of marked birds and others recognisable as individuals by distinctive plumage characters, in Teesdale, County Durham, during 1993–95 (Parish *et al.* 1997a) and near Bergen, southwest Norway, during 1991–95 (Byrkejedal *et al.* 1997) have questioned this, revealing a situation of unexpected complexity, with a significant level of polygyny.

There were strong similarities in the results. In both studies there were annual variations in the frequency of polygyny, with 20–44% of males having two or more mates over three years in Teesdale and 23–41% doing so in Norway over five years. In both studies a proportion of males on or near the boundaries of the study sites, usually yearlings, were unmated (up to 29% in Teesdale and 21% in Norway). In Teesdale the incidence of polygyny increased with the age and experience of the male. In Norway its incidence was linked to territory size, which probably reflected male quality. As part of the same study, Grønstøl (1996) found that the quality of display, particularly the vigour of the Alternating Flight (Chapter 9), predicted the number of mates a male obtained. Nevertheless both studies found that males frequently changed status in succeeding years, switching from monogamy to polygyny and vice versa. Females clearly ignored available young males, preferring secondary status with an already mated male with a good territory and/or greater experience. Cues such as quality of display flight may have indicated other aspects of male suitability in territory and nest defence (Grønstøl 1996). Most incidents of polygyny were bigamous but five males in Teesdale had three mates (trigamy) in 1995, and there were two trigamous males in 1991 and five in 1992 in Norway. Figures from both studies suggested that the frequency of polygyny was related to the number of males present. Both studies also recorded males having mates in two separate territories, two in Teesdale and one in Norway. In Teesdale Parish *et al.* (1997a) also recorded one case of a female changing mates and territories for a replacement clutch, which they considered to be the first case recorded of polyandry in Lapwings. In another study in southwest Norway in 1998 Hafsmo *et al.* (2001) recorded one male having four females simultaneously; on 19 April all four were either incubating or laying eggs.

Other recent studies have indicated similar levels of polygyny. Thus Byrkejedal (unpublished, quoted in Byrkejedal *et al.* 1997) found that 67–80% of males in another site in southwest Norway were polygynous in 1984–85, as were 29% of males in a study in Hungary (Liker & Székely 1999a). Berg (1993), studying habitat selection on Swedish farmland, found that half of 20 randomly selected territories in 1988 were held by polygynous males, with bigamy and trigamy involved. Polygynous males held larger territories and territory quality for foraging, particularly the degree of wetness, was what attracted secondary females.

It is hardly likely that these studies represent isolated instances and polygamy should reasonably be regarded as a regular mating strategy in Lapwings. Records of polygyny have a long history and the earliest convincing mention of it that I have found is in Browne (1889) for Leicestershire. Rinkel (1940) thought it to be of common occurrence, although the English translation of his summary strikes me as somewhat ambiguous; he was apparently the first observer to record trigamy. Nor does polygyny only occur in substantial groups. Wilson (1967) recorded an isolated male in Lancashire with three mates and in north Breconshire the last breeding record I obtained before the species became extinct there proved to be a solitary male with two incubating females, both of which lost their nests; other isolated cases of polygyny have recently been reported from the same county (M. F. Peers pers. comm.). In 2005, I studied a small population on the Welsh coast, comprising 11 males and ten females. Five males were monogamous, one had two females and one three; four males never attracted mates and they left the site in May (Shrubb 2005). In 2006, at the same site, there were eight males and ten females. Three males were monogamous, two had two mates and one had three; two males were unmated. At this site, until young hatched, most of the feeding that I observed occurred within territories and those of the polygynous males included substantial puddles and damp areas, where females frequently foraged. After hatching the young were often moved to neighbouring fields.

One interesting side effect of these recent discoveries about mating systems in the Lapwing is that, if widespread, which must be supposed, such behaviour means that recent population estimates based on pairs and territories may underestimate the true size of the breeding population in terms of nesting females. This probably does not affect the validity of replicable surveys for comparisons. Its effect is also reduced because significant numbers of young males hold territory without attracting mates and a significant percentage of the population breeds as solitary units. That does not necessarily preclude polygamy as the Lancashire and Breconshire records above demonstrate. Nevertheless many local populations may have more breeding females than territories.

Lapwings are normally single-brooded but two recent studies, Blomqvist & Johansson (1994) and Parish *et al.* (1997b) have recorded a total of eight cases of double brooding, where the first clutch was incubated by both partners and, after hatching, the male cared for the young and the females laid and incubated a second clutch. Parish *et al.* suggested that this behaviour might be more frequent, as it is readily overlooked or taken to refer to replacement clutches. However, in total these second brood attempts only produced one youngster, which does not argue that it is a particularly successful strategy.

PHILOPATRY AND SITE FIDELITY

The Lapwing is a strongly philopatric species, that is it has a powerful tendency to return to its natal area to breed in subsequent years. Figures given by Thompson *et al.* (1994) suggest that this may be a more marked tendency in Lapwings than in many other waders. In a study in northern England these authors found that 74% of colour ringed young birds in Teesdale, Durham returned in their first or second year to within 5km of where they were hatched and that 37% of marked youngsters did so in the Eden Valley, Lancashire. Young males and females returned to breed to their natal field, or an adjacent one, in approximately equal proportions, 45% of males and 52% of females. They also found that adults (taken to be birds three years old) were strongly site-faithful, 95% using the same or a neighbouring field for nesting in successive years, although second-year birds were somewhat less site-faithful. In Scotland Baxter & Rintoul (1953) recorded a bird ringed as a nestling in Torrance in 1915 being recovered on the same site in 1927. Nethersole-Thompson (in Bannerman 1961) noted that many females probably return to the same territory annually, which he established by the presence of distinctive eggs laid in successive years. Such a high degree of philopatry and site fidelity raises the risk of inbreeding but Thompson *et al.* found no evidence of parents breeding with offspring or siblings with siblings in their study.

The study raised the possibility that levels of philopatry may differ between local populations. Interestingly, a proportion of the Eden Valley birds bred on arable fields where crop rotation partly inhibited the strong site-fidelity shown by birds nesting in permanent grassland. Lapwings have long adapted to the habitat changes in individual fields imposed by crop rotation and they readily shift nesting- and chick-rearing fields in response (e.g. Fuller 1994, pers. obs.). It would be interesting to know whether this has a general effect on philopatry because of the greater mobility by nesting pairs required.

In Britain generally Thompson *et al.* (1994) found that 61% of Lapwings breeding in their first or subsequent years did so within 10km of their natal sites. They believed that recruitment in the Lapwing was mainly driven by philopatry and thus by breeding success in a given area. Populations decline if that breeding success declines. Philopatry therefore presumably interacts with the stability and availability of suitable habitat. While advantageous in ensuring return to a successful breeding area, it seems likely to become less so when habitat has deteriorated on the scale that has happened in modern farmland. However a significant proportion of young Lapwings disperse more widely, 28% being recovered up to 100km from their natal areas and 11% more than 100km away. Some of the latter have been found from 2,000km to over 4,000km away. Thus the species has significant potential to explore and occupy new areas (Chapter 13).

European populations show a similar high degree of philopatry, with 54% of pulli ringed in Denmark returning to less than 10km from their ringing site and 62% to within 20km (Bak & Ettrup 1982). Across Europe generally c.70% return to within 20km (Imboden 1974).

CHAPTER NINE

The breeding season: courtship, display and pair formation

Lapwings have a complex series of breeding season displays both in the air and on the ground. Broadly speaking, the aerial displays advertise, define and maintain territory and convey important information to the female in selecting a mate. Ground displays by the male attract mates and cement the pair bond. Choosing a mate is a female prerogative. Some ground displays are also territorial and there is a distinct diurnal rhythm in display activity, with most early in the morning, declining towards midday and with a resurgence in the evening. During the early stages of the breeding cycle males frequently display at night. There is also a seasonal rhythm in display flights, with most occurring during the period of territorial establishment until eggs are laid, then declining slowly and stopping when eggs hatch. Activity increases again if eggs are lost until they are replaced, or with the appearance of a new female in the territory, until she is paired or departs (Dabelsteen 1978). Poor weather depresses activity.

DEFINING TERRITORY

Dabelsteen (1978) noted that the Lapwings' display flights often followed physical boundaries such as a hedgerow or a lakeshore. In the sense that such features must place natural boundaries to a territory, this seems largely fortuitous. Competition between males appears to be the main definer of boundaries between Lapwing territories, the area of which largely rests on what the vigour of the male can hold. This was well illustrated by Berg (1993), who studied 20 randomly selected territories in mixed farmland in Sweden. Polygynous males, an important aspect of whose behaviour is the quality of their display flight (see below), held larger territories than monogamous ones and males in colonies had smaller territories than those of solitary nesters (Chapter 8). The small territories observed by Byrkjedal *et al.* (1997) (p.112) were associated with the very high density of 22 males and 36 females on *c.*6ha, again suggesting that competition limits territory size. They also noted that larger territories lay along the boundary of the site, where part of each territory did not need defending. Weak territory holders may be pushed out or leave, particularly if they fail to hold a mate, and neighbours then expand into the space created (e.g. Shrubb 2005).

Territorial establishment starts with the males leaving the flock and occupying more or less definite positions in the nesting field (Brock 1911), which form the core of the territory. These are often related to particular features in the nesting field, notably patches of rather broken ground, damp areas, flashes and pools or, in cereals, areas of sparse growth on poor soils (Chapter 10). Territory sizes vary with habitat, forms of dispersion or types of mating system (Chapter 8).

Boundary disputes are often fairly formalised. Males stand facing one another, standing very upright and puffing out their breast feathers to show off the gorget. The wings are often drooped a little at the shoulder and they will turn and walk parallel thus along an 'imaginary' boundary, or spin round and present their backs, or make feeding movements before again staring at one another. When territories are well established they may do little more, before relaxing and moving back into their territories but such 'staring matches' can last for half an hour or longer. Earlier in the season, however, they often launch into what I term fluttering up fights, facing each other and flying up with legs dangling and striking at each other with their wings, each trying to rise above the other. Spencer (1953) noted that three or four males may participate in such flights and birds may go well outside their territories to join in, attracted by the activity. Such flights often escalate into aerial chasing, similar in form to the alternating flight of the display (see below). Such chases also include parallel diving. After a contest males will dive back into their own territories and perform song flight displays.

I have not found fighting on the ground by Lapwings to be very common but it may occur between males, between females and occasionally between males and females. Selous (1933) noted that each fighting male tried 'to spring above the other and to strike down upon him, and there are not many such buffetings before it becomes a rough and tumble and, one seizing the other by the neck with his beak, both roll together on the ground, the bird that has the grip endeavouring to hold his opponent down whilst continuing to strike him with his wings'. Although also noting that ground fighting was uncommon, Spencer (1953) observed rather similar behaviour, birds springing together and buffeting each other, with intervals of standing erect with wings partly unfurled in the high upright posture. Haviland (1915) noted that 'sometimes two birds charged each other on the ground with outspread wings, but when within striking distance each twirled round and stood with open wings with his back to his rival and bill drawn back'. Males also run at one another in the threat attitude as described for winter territories (p.100): stiff-legged, head tucked in, bill level, tail held rather above the horizontal and the back feathers often ruffled. There seem to be many variants around these themes. In particular Nethersole-Thompson & Nethersole-Thompson (1986) describe an interesting leap-frogging display, which seems somewhat to resemble Haviland's observations above.

Liker & Székely (1997) measured territorial aggression between females on their grassland site in Hungary, presenting dummies made from taxidermist's mounts of male and female Lapwings near active nests. They found that resident females attacked female dummies more often than male ones. Attacks on female dummies were mainly on the ground but males were attacked equally on the ground and by aerial dives. Attacking females used typical territorial postures and displays. They also pecked the dummies or hit them with their wings or kicked them. When diving attacks were made they also sometimes struck the dummy with their wings. Such territorial aggression towards females declined as incubation progressed but not aggression towards males. Liker & Székely proposed that this behaviour by resident female Lapwings aimed to prevent their mates from acquiring extra females and so to monopolise the male's parental care for their chicks.

TERRITORY: THE DISPLAY FLIGHT

The most important display in advertising, marking and defending territory, however, is the display and song flight. Dabelsteen (1978) divided it into six different phases, a scheme I have followed. The display starts with the butterfly flight, which I prefer to call the owl flight, because it invariably reminds me of a Short-eared Owl. The male Lapwing rises from the ground with 5–10 very deep and slow wing beats with expanded flight surfaces. This flight may often be performed on its own, particularly as males fly to females to copulate; if she refuses the male may briefly adopt the high upright posture before pattering away.

From the owl flight the male quickens into normal flight and then goes into the alternating flight, in which he flies on a rapid zigzag course, throwing himself from side to side, conspicuously flashing alternate black and white, in a roughly circular or oval course round the extent of his territory. The flight is accompanied by a loud thrumming note caused by the three outermost primaries, and to some degree the fourth, being made to vibrate on the downstroke (Dabelsteen 1978). It thus has an affinity to the drumming display of the Common Snipe. Grønstøl (1996) noted that the frequency, amplitude and speed of the wing beats in this flight are greater than in ordinary flight and that the angle of roll measured on video stills averaged 52.3° with a range of 20–90°. When the bird adopts extreme angles (>85°) the wing beats become assymetrical, with the upper-wings beating at a higher amplitude than the lower and males may come off high roll angles into a dive. As with the owl flight, the alternating flight may be performed on its own, particularly prior to coition or in flights against neighbouring males.

The alternating flight is followed by the song-flight proper. I have rendered the Lapwing's song as a ringing 'pee-wip, wip wip, pee-a-wip', each phrase accompanying a different stage of the song flight, the 'pee-wip' as the male climbs steeply out of the alternating flight, the 'wip wip' as he rolls off the top of his climb and the climactic 'pee-a-wip' in a steep tumbling dive in which he often goes right over onto his back, a beautiful display and song. The song flight is then followed by another bout of alternating flight and the male may perform several sequences of song and alternating flight before landing.

Dabelsteen (1978) noted that *c.*20% of display flights occurred without external stimulus but that most (60%) occurred in response to the presence of another male, either flying by or also displaying or engaging in territorial activity. If such males obtrude too closely the territory owner will break off display and give chase. Males also respond by display to the presence of females, which they may also chase, apparently trying to push them down onto their territory (pers. obs.); such chases may be extensive and vigorous and I suspect may be instigated by the female. Display flights also occur in response to predators, which will be attacked if they come too close and, in the early season, males display on return to their territories from neutral feeding grounds.

The Lapwing's display flight serves three functions. It advertises the presence of a male with a territory, it defines the territory's extent and, because the flight is technically and physically demanding (Grønstøl 1996), it advertises the fitness and eligibility of the male, his ability to maintain a good territory and to keep out potential rivals and predators. Elements of the display flight, particularly the alternating flight and the steep dive with the climax of the song, closely resemble attack procedures against rival males and predators.

Grønstøl (1996) analysed the display flight in detail, using video recordings of displaying males on a site in southwest Norway with a high density of breeding Lapwings (see also Byrkejedal *et al.* 1997). Grønstøl's observations showed that the alternating flight, which comprised about 60% of the song flight, was the most important component influencing the female's selection of a mate. He found that steepness of the angle of the roll in alternating flight correlated with mating status in males, polygynous males having more vigorous flights with greater angles of roll. Laying dates correlated with the same factor which also, to some degree, predicted territory quality, as it correlated with the average biomass of earthworms on territories. There was a closer relationship between mating status in males and the alternating flight than with factors such as territory size.

Essentially Grønstøl's study indicated that female Lapwings have a strong predilection for males which are good aerobatic flyers. He noted that confrontations between males occur frequently and often go into flight chases similar in manner to alternating flight; as noted above such flights also include parallel diving. After such contests males often perform song-flight displays, which are also induced by other males passing over the territory. This provides females with many opportunities to assess and compare males' abilities in defending territories.

These results have been supported by studies in Hungary by Liker & Székely (1999a), who found that, although territory size or the abundance of food on males' territories did not influence males' mating patterns, males which became polygynous spent more time in acrobatic aerial displays than those which remained monogamous.

Variations in display performances may occur, particularly in the early stages of the breeding cycle if a party of females flies over the territories. This may trigger a wild display of aerobatics by the males present. Although they start as formal display flights, they develop into virtuoso performances, with high speed diving and twisting and climbing and zigzagging with complete abandon, more nearly resembling Spencer's term 'crazy flying' (Spencer 1953) than any formal display I have observed.

GROUND DISPLAYS: CHOOSING A MATE

The main ground displays between the sexes centre on the nest scraping display, in which both sexes ultimately participate. However, when they start to move on to and investigate the males' territories, female Lapwings are remarkably unobtrusive. Nethersole-Thompson & Nethersole-Thompson (1986) used the very apt word 'furtive' to describe their behaviour at this stage. A female will often stand quietly and watch the male's scraping display and may then simply fly away or she may move slowly towards him, pausing at intervals, apparently feeding. The male may walk or trot towards her at this stage, using the rasping or scratchy 'zree zree' mating call, puffing out and displaying his black breast band before pattering away, with the breast band still expanded and wagging his tail at intervals, back to his scrape, to rock vigorously interspersed with spasms of grass throwing (see below).

In the scraping or rocking displays the male always presents his tail to the female. He tilts forward onto his breast and scrapes a hollow with his feet, moving from side to side as he does so in a movement which gives the appearance of shaping the hollow with his breast. The tail is vigorously wagged, alternately showing the orange undertail coverts and the white rump area. Under greater excitement the wing-tips are raised so that they point straight up into the air and the whole rump and tail is exposed. At times he also jerks bits of grass over his shoulder and along his flanks. If the female shows no interest and moves off, display will stop and the male may do a flight display or move off to feed. Females may visit several territories in this way. As noted under philopatry (page 121), however, many females may well return annually to the same territory. Under these circumstances it is likely that they often re-establish an old pair bond, since males and females show similar high levels of site fidelity.

Eventually, however, the female moves slowly towards the displaying male and stands beside or behind him. The male then steps out of the scrape and bows steeply, presenting and displaying the orange undertail coverts fully to her. The female steps into the scrape and also rocks, similarly if less vigorously than the male. The male may displace her and scrape again, while she stands by, picking up and flicking bits of grass or other material over her shoulder. Coition may follow these displays or the female may simply move away a little. My experience is that once these exchanges have taken place, the female has usually made a choice, although this is not invariable, and seductions occur. This display behaviour is illustrated in Figure 9.1.

Coition does not have any obvious signal that I have observed but Nethersole-Thompson describes the female *'dipping stiffly forward on taut legs'* to signal her readiness. The male often flies to the female in the owl flight, sometimes from a distance, and mounts directly from flight, waving his wings to keep his balance. This seems much more frequent behaviour than mounting from the ground, when the male runs to the female in the hunched posture giving the scratching call. Coition continues right up until hatching and particularly may occur when the birds change over at the nest. Females also sometimes leave their nests to mate and then run or fly back to the nest (Nethersole-Thompson & Nethersole-Thompson 1986). Extra-pair copulations are not unusual in Lapwings and may be instigated by either sex.

Male Lapwings usually have several scraping hollows and may have as many as 20; they are often nest scrapes formed in previous years (Cramp & Simmons 1982). Display continues after a pair has formed and eggs have been laid, the male attempting to attract other females to himself. A male without a female will also perform scraping displays, advertising his readiness to mate and trying to entice one, a process that may take two or three weeks. If a nest is lost the male starts another cycle of scraping displays.

Accounts in the literature draw attention to many variations in this display behaviour, perhaps largely the result of the pressure of excitement or frustration. But the central themes outlined here are constant.

Figure 9.1. A scraping display sequence. (i) male scraping, tail to female, displaying undertail-coverts and white rump area. (ii) female approaches and male out of scrape, bowing steeply and displaying undertail-coverts. (iii) female into scrape and scraping; male flicking material along flank.

DISPLACEMENT AND DISTRACTION

The ritualised habit Lapwings have of plucking grass or picking up pieces of vegetation or small pebbles and flicking them along their flanks or over their shoulders is not confined to scraping displays. It occurs often when either sex is emotionally stressed, excited or frustrated and seems clearly then to be a displacement activity to release such stress. Females may also do it when they are relieved at the nest and it may occur in other situations. One female I watched, whose male had acquired a further mate, was much disturbed by the new female's presence nearby when that bird started to incubate, continually hopping off her nest, stumping around and throwing grass vigorously before flying off to feed near the male, who flew on to the eggs. She stayed off the nest for well over an hour, alternately feeding and throwing grass before finally returning and settling down. By the following morning all was calm but her behaviour this first morning suggested nothing so much as temper.

Nethersole-Thompson & Nethersole-Thompson (1986) also noted that agitated Lapwings stamp their feet or stab the herbage, actions which resemble feeding or attack procedures. Birds also false-feed, tipping forward in feeding motions but with the bill closed and taking nothing. Females seem to do this when moving slowly towards a male displaying on his scrape.

Distraction displays which simulate injury seem to be uncommon in Lapwings. Lynes (1910) provided a detailed description of a Lapwing decoying a Stoat away from its chicks, flapping on the ground as if wounded and spreading its wings as if unable to fly, screaming all the while. It kept just out of reach of the Stoat, slackening its speed when the Stoat went more slowly and enticed it 400m from the brood. Where distraction displays directed at potential predators have been described in the literature they seem to conform fairly closely to Lynes' description but Coward & Oldham (1904) recorded an incident where the female 'jumped spasmodically and then pitched forward on its head, waving its wings disjointedly and struggling along the ground as if it were badly hurt. Then it lamely struggled down the ridge towards the channel of a stream, entered the water and swam leisurely across, holding one wing elevated . . .. Reaching the far bank, it walked slowly up the mud, every few yards repeating the epileptic leaps and falls . . .' before flying off.

In his experiments on the anti-predator responses of Lapwings (Chapter 11), Elliot (1985b) found that, as his dummy fox got close to the nest, adults attempted to lead it away, making short runs to the side and past the nest and back. As the fox advanced the intensity of this behaviour increased leading into distraction displays, flapping wings back and forth, showing alternate black and white and showing the brightly coloured tail. Birds often glanced back or turned to face the fox during this performance and sometimes called throughout. Only males were observed to perform these distraction displays.

The banner-waving display used to divert large herbivores from the nest is described on p.159 but Spencer (1953) described an incident in Orkney where a female with one tiny chick displayed similarly to him and a companion, running up to them calling distractedly and waving one or both wings. Now and then she took wing and flew just overhead with legs dangling, then alighting to display again. Occasionally she keeled over on one side as if her legs were weak. In momentary intervals between demonstrations she displacement-fed, very hurriedly.

The main reaction to human beings in the vicinity of chicks is similar to that described for foxes. If you walk through a rearing field male Lapwings will often move to stand prominently and watch and assess your approach before taking off. The birds then fly to meet the intruder with a distinctive shallow wingbeat, calling continuously a plaintive 'peee-wi' with the accent on the first syllable. They circle calling all the time the intruder is in the vicinity and this characteristic flight and call is clear evidence that young are present. Other observers have noted that males may dive at intruders more closely in typical mobbing actions but I have found this to be distinctly rare behaviour. Interestingly, however, reading 19th century accounts leaves the strong impression that it was once more usual. For example MacGillivray (1837–52) describes how both birds of the pair 'fly about, now high, now low, suddenly descending and rising in gentle curves or abrupt windings, and performing a variety of evolutions, sometimes striking their wings so forcibly as to cause a loud noise, and usually emitting their peevish wail. So great is their anxiety, that they will frequently come very near . . .'. Other 19th century avifaunas contain rather similar accounts but whether they were typical or were recorded because they were unusual is often not made clear.

REACTIONS TO OTHER SPECIES

Although sharing their breeding sites with species such as Snipe, Curlew and Redshank, Lapwings are often hostile to other waders, both passage and breeding birds, trespassing within their territories, perhaps particularly to Golden Plovers and Curlews. Nevertheless other ground-nesting species take advantage of the effectiveness of Lapwings' anti-predator behaviour to reduce their rates of nest predation by nesting close to them. Such associations have been recorded for Snipe, Redshank, Ruff, Meadow Pipit and Yellow Wagtail. Thus Mayo (1974) recorded Redshanks nesting within a few metres of Lapwings' nests and Snipe as close as four metres. Kvaerne (1973) observed that at four out of five Lapwing nests he found on Dovrefjell, Norway, in 1972 there was a Redshank's nest within 7–12 metres; the Redshanks benefited from the early warning of the approach of predators provided by the Lapwings. Eriksson & Götmark (1982) found similar associations between nesting Lapwings and Meadow Pipits and Yellow Wagtails: the passerines occurred at higher densities and had lower rates of nest predation within Lapwing territories than without. Dyrcz *et al.* (1981) noted similar associations between Redshanks, Ruff and Snipe and Black-tailed Godwits in the Biebrza marshes

Poland, the other waders benefiting from the defensive umbrella provided by active defence by the godwits.

Other species arouse Lapwings' ire. Some of these, such as Grey Herons, are predatory but others are not. Cock Pheasants are especially likely to be set upon (see frontispiece) and I have watched the entire population of a field holding 12 pairs of Lapwings queueing up to dive bomb one that had wandered into their midst; the pheasant eventually fled, ducking and weaving as he did so. I have never worked out what causes this reaction by Lapwings, since there seems to be no obvious reason why gamebirds (partridges may be similarly treated) should be regarded as any threat. Many other species are totally ignored and this includes wildfowl such as Canada Geese. Nor have I noted any reaction to mammals such as rabbits or hares, which often occur in nesting fields.

AGE AT FIRST BREEDING

The age at which Lapwings first breed seems to be subject to considerable variation. Nethersole-Thompson & Nethersole-Thompson (1986) considered that many only bred for the first time in their third calendar year. Age provides experience and, for males, may well improve the power and quality of the display flight, which would improve the chances of obtaining and holding a mate. That this is so is suggested by the fact that three-year-old birds are much more likely to mate polygynously than yearlings (32% *cf.* 13%, Parish *et al.* 1997a).

Nevertheless Thompson *et al.* (1994), in their study in northern England during 1990–92, found that 67% of male and 69% of female Lapwings hatched in 1990 bred in their first year, whilst 29% of males and 25% of females hatched then bred for the first time at two years old and 6% of both sexes at three years. The study was conducted over two sites, in Teesdale and the Eden Valley, and the proportion breeding in their first year differed, with a much lower proportion doing so in Teesdale (68%) than in the Eden Valley (90%) where the population was declining quite sharply; that in Teesdale was stable. Even on the Teesdale site there seemed to considerable annual variation in the proportion of yearlings breeding. Parish *et al.* (1997a), studying mating systems there during 1993–95 found that 67% of one-year-old males were unmated, compared to 33% in 1990, whilst 85–90% of males had mates at two years old or older. Factors such as changing age structure in local populations or local population declines may well give younger birds better chances of obtaining and holding mates than are available in settled stable communities with their high level of site fidelity.

Males which do not attract a mate nevertheless obtain and hold territories. These may be on the fringes of the breeding area in less favourable habitat and Glutz von Blotzheim *et al.* (1975) also noted that yearlings were often late arriving in Switzerland, which also meant that they occupied poorer territories.

CHAPTER TEN

The breeding season: laying, incubation and hatching

In farmland, Lapwings not only select particular types of field for nesting, they may choose consistently to site their nests in particular parts of those fields. For example on tilled land on our family farm in the past, nests were not only placed on arable fields abutting pastures (Chapter 8), they were almost invariably placed in parts of those fields which were remnants of old beaches or shorelines around what had once been tidal creeks. The soil in these places was poor and stony, limiting crop growth and incidentally giving good cryptic protection to the nests and eggs. However, with little general evidence of a preference for particular soil types for nesting, choice was most likely to be based on the poor crop growth and, as in grassland, crop growth is an important determinant of where Lapwings nest in tilled land. Bare land or the early stages of crop growth are preferred and water-damaged stands of arable crops they would otherwise avoid, such as autumn cereals, are also used.

In grassland Klomp (1954) observed that females apparently select fields by colour. He postulated that the grey-green or brown-green colour of favoured fields was selected because it indicated delayed growth (see Plate 20). Interestingly, the colour of the female's upperparts matches this colour well, which may be a contributory reason. Grassland nests are often also sited in broken ground, for example in areas of variable sward height and plant composition which make patches of different densities and colour, in patches of tussocky rough grassland or in wet spots poached by cattle in winter, where the combination of brown-green colour and a broken background renders the sitting female remarkably inconspicuous. Seebohm (1885) noted that the hoof prints of cattle or horses were often selected as nest sites and Nethersole-Thompson & Nethersole-Thompson (1986) recorded Lapwings using the old nest scrapes of Golden Plovers. The vicinity of flashes and puddles may also be chosen and Lapwings also select sites among clumps of soft rush mown in early spring, which provides another form of broken background, without inhibiting the sitting birds' field of view.

Although most Lapwings nest on farm fields or in open agricultural grasslands, a wide range of habitats outside farmland is used regularly (summarised in Table 4.3). In general, however, most of the latter nest sites seem to provide the basic requirements of a cryptic and/or broken background, providing concealment for the nest and incubating female.

As already noted, females avoid nesting in the vicinity of trees because of their use as hunting perches by predators. Nests are therefore placed away from field boundaries and there is a tendency to avoid small fields, which are most likely to be occupied in areas of high population density. Birds may be more tolerant of boundaries such as fences, which do not impede their field of view, despite their utility as perches for crows. As an example five nests, one a repeat clutch, I watched recently in Wales were each placed within 10–20m of fence posts, although the males' territories would have allowed greater distances. These nests were in patches of broken ground, providing good cryptic cover, which no doubt governed choice. Nests also tend to be sited nearer good feeding or chick-rearing sites, tillage abutting pasture for example, or closer to streams and flooded areas (Berg *et al.* 1992) or to rills or damp areas in grass marsh (Milsom *et al.* 2002). Reijnen *et al.* (1996) also found that road traffic reduced nesting density in Dutch grasslands, the effect declining with distance from the road and varying with the weight of traffic. At very high traffic densities reductions of as much as *c.*50% occurred at 100m, *c.*37% at 500m and *c.*16% at 1,500m from the road. Farmsteads and buildings also tend to be avoided and Milsom *et al.* (2000) found that territories tended to avoid overhead power lines, which might be because the cables could inhibit display flights.

However, the overriding requirement for nest-site selection is short vegetation or bare ground, providing a clear field of view for the detection of predators and other hazards to the nest. Klomp (1954) examined this in some detail and noted that in dense herbage vegetation heights of more than 4–7cm were avoided but greater height will be tolerated as the vegetation thins: for example, up to 12–15cm in winter cereals sown at 15cm spacings. Hart *et al.* (2002) found that the optimum sward height at territorial establishment in the Kent grazing marshes was 3–4cm. In ungrazed areas nesting frequency declined in May as sward height increased to over 10cm. In grazed marshes, where sward height remained at about the optimum level, the frequency of nesting attempts peaked in early May. In Öland, Sweden, Ottvall (2005) found that the mean vegetation height at nests was 2cm and, in alkaline grassland in Hungary, Liker (1992a) found that it was typically 5–12cm.

Much higher vegetation may be used at times where open patches occur in areas of soft rush or Yellow Flag and incubating birds in such sites may have a fairly restricted field of view. Imboden (1970) noted that, by the time of hatching, nests in Swiss grassland were often in vegetation 40–45cm high. In arable crops the effect of increased height and density in autumn cereal crops early in the season was discussed in Chapter 4 but H. Matter, quoted by Glutz von Blotzheim (1975), recorded nests in oil-seed rape up to 100cm high, with a mean of 54cm for 15 nests. Drill spacings were apparently not reported but may be important in influencing site selection in arable crops. Maize crops provide a good example. Normally sown at 40–50cm spacings, it still provides satisfactory conditions in June (Plate 18). I once found a nest in waist-high wheat, admittedly on a patch killed by a winter puddle. The eggs hatched successfully and the female took the young 100m along a tramline to a neighbouring pasture.

Some extremely eccentric sites have been recorded recently. The BTO's nest record cards include nests found in 1975 and 1976 on a traffic island on the A435 in Warwickshire and Holland *et al.* (1984) reported a similar site in Greater Manchester. These authors also report 1–2 pairs successfully using a flat factory roof in Stockport in 1981–83 as well as nests on recreation grounds and inner city rough ground. Williams (1999) reported a further case of rooftop nesting on a school in north Wales and Calbrade *et al.* (2001) recorded two pairs nesting on adjacent sloping corrugated roofs in central Rochdale during 1998–2000. The young were reared in the trough between spans, where enough invertebrate food presumably accumulated. These authors also recorded territorial activity at rooftop sites in Walken and Ashton-in-Makerfield, where the rooftops could not be examined and that, in 1999, pairs bred on lawns at warehouse complexes in Bolton and Chadderton, the roofs of which were used for roosting. The nest record cards also recorded two nests in gardens and six on urban playing fields. Cramp & Simmons (1982) quote one other case of roof nesting in Germany and Beintema (1986) recorded similar behaviour in Oystercatchers and Little Ringed Plovers in The Netherlands and in Killdeer in North America. Such adaptations perhaps indicate increasing difficulty in finding suitable natural or farmland sites. Certainly the many early avifaunas I have consulted recorded no comparable nests for Lapwings, nor did Spencer (1953).

For the nest females normally select one of the scrapes made by the male in his rocking displays. Spencer (1953) describes females apparently selecting nest sites, usually rather early in the morning 'tripping along like Redshanks, searching about restlessly just as if they had lost something'. He suspected that they were going from scrape to scrape. Once a selection is made both sexes help to enlarge and line the scrape, for which Glutz von Blotzheim (1975) gave a mean diameter of 12cm and depth of 5cm in a sample of 26 nests. Lining material is usually what is to hand: grasses, hay, stubble straw, heather and willow twigs, moss, dead ragwort stems, bits of wood, occasionally pine needles and so forth. The female continually adds bits to the nest and will reline it if necessary. Where the nest is near water she may build it up considerably if flooding threatens, although not always successfully.

TIMING OF LAYING

As with arrival and territory establishment, laying in Lapwings starts later as one moves north and east across the range. Figure 10.1 illustrates the broad geographic progression of laying recorded across Europe and western Asia. The figure shows the period when laying is recorded as starting in each time band. The data for Russia and western Siberia are drawn from Dementiev & Gladkov (1969), so may not fully reflect the range expansion that has occurred there since (but see Chapter 13), but that seems unlikely to affect the general pattern shown.

Within any population or group, however, laying dates show a different pattern as females select mates and settle. Studies in Britain provide clear examples. Thus Jackson & Jackson (1975) showed the laying of first clutches on their New Forest site extending from the beginning of April to the first week in June. Such very late

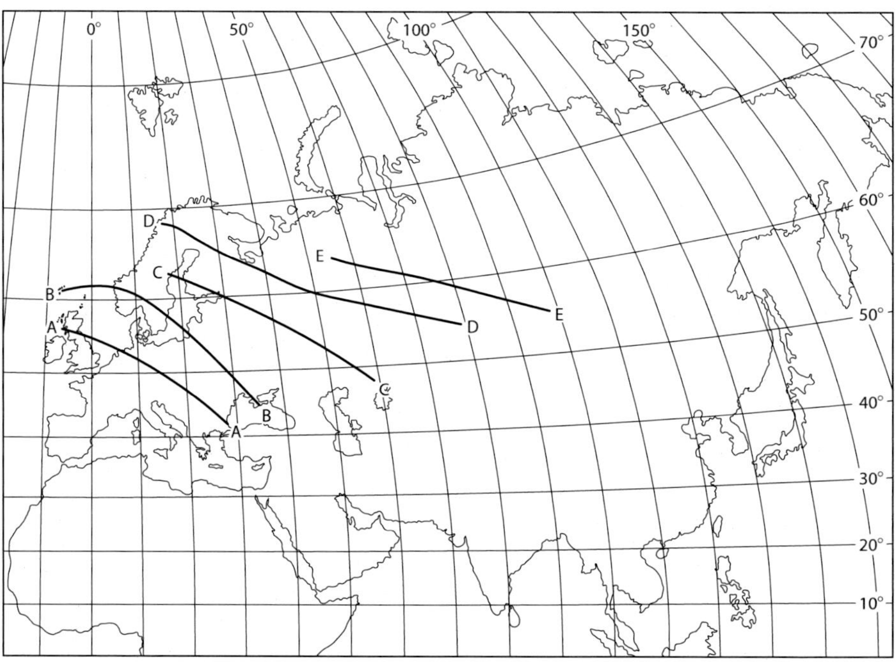

Figure 10.1. Progression of laying in Lapwings across Europe and west Siberia. Time zones are: A–A, second half of March; B–B, early to mid-April, C–C, late April to early May, D–D, early to mid-May, E-E, mid-May to June.

first clutches, however, may be open to the suspicion that the females involved had failed elsewhere and moved for another attempt. In my Welsh population in 2005 and 2006, nesting on old pasture, females starting incubating first clutches on 23 and 27 March, 1 April (3 clutches), 2, 3, 8, 9, 12, 13, 18, 20, 24 and 26 April, 1 and 4 May, and two about 6 May, with a median date of 12 April; there was no difference in timing in the two years. This seems a fairly typical pattern in Britain and Nethersole-Thompson & Nethersole-Thompson (1986) gave a series of records for northern Scotland up to 1979 with mean dates for clutch completion ranging from 15 to 24 April; laying dates there were possibly earlier in the 1940s than the 1970s. In northern England they quoted records from Arthur Whitaker's diaries with completion dates for 175 clutches ranging from 24 March to 9 May, with a mean of 19 April; 25 clutches in East Anglia from the same source were rather earlier, with a mean date of 14 April.

The timing of laying may also vary with habitat. Chamberlain & Crick (2003) showed mean first egg dates in the UK during 1990–1999 ranging from 12/13 April on pasture, to 16 April on bare land and between 19 and 29 April in different arable crops. Laying dates also vary between females on the same site according to the food supply within the territory (Högstedt 1974) and between seasons with spring weather conditions, particularly temperature. Thus Imboden (1974) showed a very strong correlation between the start of breeding in 14 populations across Europe and mean minimum temperatures in April. In Denmark Ettrup & Bak (1985) noted that egg-laying started 5–8 days later in years when the average March temperature was 1°C below normal compared to years when it was 1°C above normal. Laying was also about eight days later north of 57°N compared to the area south of 56°N. In The Netherlands Kruk *et al.* (1996) showed that median first egg dates of Lapwings, Redshanks and Black-tailed Godwits varied annually in line with the T-sum calculations which farmers use to time operations such as fertiliser applications and which subsequently influence mowing dates. T-sums are simply the sum of the maximum positive temperature each day from 1 January and effectively measure increasing soil temperatures. These not only influence farming decisions but also factors such as the abundance of prey available to Lapwings.

Laying intervals also vary between females, some laying daily, some producing a clutch of four eggs over five or six days and some laying every other day. Longer laying intervals have been noted and laying intervals are not necessarily uniform within a clutch. For example, Chance (1930) recorded an interval of at least eight days between the laying of the second and third eggs in a Berkshire nest. As a female approaches laying she exhibits distinct signs of broodiness, feeding or standing quietly in the vicinity of the nest and doing very little for much of the time. During laying she tends to remain, often very inconspicuously, quietly feeding in the vicinity of the nest. This behaviour rather resembles that of raptors, species which also lay rather large eggs, in the same circumstances.

In Britain the earliest nest record that I have found is for Lancashire, Oakes (1953) noting a nest with three eggs at Brindle, near Hoghton, on 25 February 1882. In the same county Mitchell (1885) recorded a nest with one egg at Clitheroe on 1 March 1883 and Spencer (1953) gives several nests in Ayrshire on 2 March 1921. In a detailed examination of the BTO nest record cards for England and Wales from 1962 until 1985 5% of pairs had completed clutches between 15 and 31 March, 33% between 1 and 14 April, 28% between 15 and 30 April, 18% between 1 and 14 May and 16% after 15 May. Nesting was earlier in the south and earlier in lowland grassland than upland (Shrubb 1990). There was also a consistent trend for there to be a longer laying period in tilled land than in grass and in habitats outside farmland than in either farmland category (Figure 10.2).

Individual studies may show variations from this overall picture. Thus Galbraith (1988a) found laying in May was more frequent in his rough grazing site than in arable land because crop growth curtailed the laying season in spring cereals. In contrast, on our family farm in the 1950s and early 1960s I used quite regularly to find nests on the point of hatching in late June or early July on land being cultivated for roots, particularly turnips, which were preferred nesting habitats (Figure 8.1). Such crops provided bare land for nesting throughout the breeding season but regular cultivations at intervals from late March meant frequent nest loss. We stopped growing turnips in 1968 and, in Britain generally, such crops have declined in area and distribution almost continuously since the 1950s. Thus, over time, changes in arable rotations may have influenced laying patterns in Lapwings nesting in tilled land more generally.

Chamberlain & Crick (2003) found no significant trend in first egg dates in the UK during 1962–99. However, some shift in the timing of nesting was apparent in England and Wales over a longer time scale, with a trend to earlier nesting in grass and later nesting in tillage after 1962 compared with 1940–1961 (Shrubb 1990). The small shift to later nesting I detected in tilled land after 1962 seemed to reflect the impact of higher March rainfall, particularly during 1974–85, on the timing of

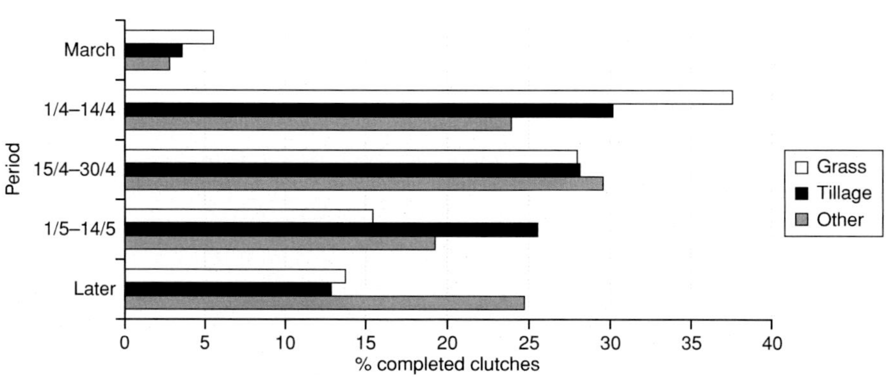

Figure 10.2. The percentage of completed clutches laid by Lapwings in different periods during the breeding season in three habitat categories in England and Wales during 1962–85. Other = habitats outside farmland. There were significantly more later clutches in tillage than in grassland ($\chi^2/_1$ = 47.36, $p<0.001$) and more in habitats outside farmland than in either (grassland $\chi^2/_1$ = 72.96, $p<0.001$, tillage $\chi^2/_1$ = 39.77, $p<0.001$). Source: BTO nest record cards analysed for Shrubb (1990).

spring cereal cultivations. These were consequently delayed into the Lapwings' peak laying period, leading to higher nest losses and more frequent replacement clutches (Shrubb 1990).

Beintema *et al.* (1985) recorded a marked shift in the timing of breeding in The Netherlands over the 20th century. Breeding shifted forward by two to three weeks from 1911 to 1974. This resulted from changes in grassland management, particularly improved drainage and higher rates of fertiliser usage, which promote earlier growth of grass and thus earlier grazing and mowing. In England and Wales the pattern has been rather different, since there is little evidence since 1940 of an overall change in the time that laying begins. Rather, the pattern shown in grassland by the nest record cards was of a significant decline after 1961 in the number of clutches found after 14 April (Shrubb 1990). This suggested the possibility of a decline in the rate of clutch replacement in grass compared to either tillage or habitats outside farmland which, as in The Netherlands, was associated with increasingly intensive grassland management (Shrubb 1990). The difference seems likely to be rooted in the unique nature of the Dutch meadow bird habitats (see Beintema 1981).

Beintema *et al.* (1985) concluded that the shift in timing they observed was the result of agricultural intensification. This finding was supported by similar analyses performed for Denmark for the period 1924–1978 by Ettrup & Bak (1985), and for Tipperne nature reserve there for 1928–1992 (Thorup 1998). They did not find a similar pattern to that recorded for The Netherlands because Danish Lapwings, breeding primarily in arable land, saltmarshes and other wet grassland (Appendix 2) rather than on intensively managed agricultural grasslands, have not been exposed to the same pattern of agricultural change.

Byrkjedal *et al.* (1997) found very little synchrony in the timing of laying by females in polygynous pairings. Secondary females began laying at times varying from the same day to three weeks later than primary females. Tertiary females laid up to four weeks later. In both cases synchrony increased with later laying. Such variations presumably reflect patterns in the acquisition of additional females, which certainly seemed to be the case with birds I watched in Wales in 2005 and 2006. Byrkjedal *et al.* (1997) also noted that the primary females of polygynous males tended to lay earlier than monogamous females, which had similar egg-laying dates to secondary females.

From Britain right across Europe to Belarus the Lapwings' laying season extends to around mid-June, which means that the length of the laying period declines from west to east, as breeding birds arrive later. Dementiev & Gladkov (1969) gave more information for fledging times than about laying seasons but right across the former Soviet Union young were noted fledging from early to mid-June in the west until July as far east as Lake Baikal. Thus, again, laying in these regions was unlikely to be later than mid- to late June. Imboden (1974) noted a similar phenomenon in northern compared to central European populations, with laying starting later in the north but finishing at about the same time. Very occasionally much later clutches are noted. Wilson (1926) found a juvenile in Inverness which he judged to be only 10 days old on 1 September 1926. The clutch from which it derived cannot have been laid earlier than the third week in July.

CLUTCH SIZE

The Lapwing is a determinate layer of four eggs (Lack 1947). Experiments have shown that, if eggs are removed during the course of laying, the female will continue laying until a clutch of four is produced, sometimes in a new nest (Rinkel 1940, Klomp 1951). Nevertheless clutches of less than four are quite common in all populations, perhaps particularly in areas of poor habitat (e.g. Bolam 1912), and larger clutches are not particularly rare, with about 1% of the total clutches listed in Table 10.1 being of five eggs. Kelsall & Munn (1905) and Nethersole-Thompson & Nethersole-Thompson (1986) both provided evidence, based on unusual eggs, that laying clutches of five may be an hereditary trait. Clutches of six to eight have been recorded but are always open to the suspicion that two females were involved.

Table 10.1 summarises the clutch size recorded in a number of European studies. The table shows rather little variation over much of the European range and, in those studies where clutches of different size were shown, 1% of these clutches were of one egg, 4% of two, 14% of three, 80% of four and 1% of five eggs. Clutches in grassland in England and Wales were significantly smaller, however, a pattern common to rough grazing and improved grass and to uplands and lowlands; clutch size in habitats outside farmland there was similarly small.

The BTO nest record cards show little evidence of any marked long-term trend in clutch size in England and Wales, at least since 1950 (Figure 10.3). Clutch size in tilled land tends to be larger than in grass and that trend holds for the UK generally and until 1999 (Table 10.1. Chamberlain & Crick 2003) but the figure does show a fairly clear trough in clutch size between the mid-1950s and mid-1960s, coinciding with the peak of the organochlorine pesticide era. It is most marked in arable land nests and suggests that breeding capacity in Lapwings was affected, as was that of many birds (Newton 1998). The decline in this period is supported by data on clutch sizes gathered before 1951 (Table 10.2).

Chamberlain & Crick (2003), however, noted a marginal but statistically significant increase in clutch size over all habitat categories during 1962–99 for the UK. They noted that increased clutch sizes are a phenomenon which has frequently been observed with declining populations of farmland birds in Britain but its scale in Lapwings seemed too small to have much biological significance. However, it seems likely that the increase they observed at least partly reflects a slow recovery from the low reached during 1950s and, by the 1990s, they still could not show that overall mean clutch size had recovered to the level of the pre-1952 era in England and Wales. The pattern in mean clutch sizes noted in Table 10.1 suggests that Lapwings in Britain now lay fewer eggs than those in continental Europe.

Table 10.1. Mean clutch size in Lapwings in Europe.

Country and habitat	*Number of nests*	*Mean clutch size*	*Standard deviation*	*Reference*
England & Wales 1962–85				
Grassland	2,269	3.63	0.70	Shrubb 1990*
Tillage	844	3.78	0.54	Shrubb 1990*
Non-farm habitats	635	3.63	0.68	Shrubb 1990*
N. England 1985–1987				
Grassland	575	3.62		
Arable land	74	3.74		Baines 1989
Wales 1995–96				
Reclaimed opencast coalmine	135	3.73	0.623	Dixon 1995, 1996*
Scotland				
Arable land	85	3.75	0.55	
Rough grazing	121	3.74	0.51	Galbraith 1988a*
Upland pastoral	202	3.93	0.44	Nethersole-Thompson & Nethersole-Thompson 1986*
France	815	3.80		Trolliet 2003
Belgium				
Farmland	163	3.89	0.399	Van Impe 1988*
Uncultivated land	249	3.91	0.298	Van Impe 1988*
The Netherlands				
Grassland	57	3.88	0.378	Klomp 1951*
Denmark	132	3.83	0.484	Ettrup & Bak 1985*
Tipperne N.R.		3.88–3.91		Thorup 1998
Germany	89	3.69	0.646	Ettrup & Bak 1985*
Mixed farmland	197	3.86	0.415	Kooiker 1990*
Arable/grassland	244	3.70		Matter 1982
Finland	323	3.89	0.406	Heim 1974*
Latvia	207	3.73		Trolliet 2003
Switzerland	558	3.85	0.509	Heim 1974*
Arable land	1,395	3.81		Matter 1982
Hungary				
Grassland	128	3.87	0.403	Liker 1992a*
	67	3.89	0.305	Hegyi & Sasvári 1998
Ukraine	66	3.74		Trolliet 2003
	27	3.89		Trolliet 2003

References marked * provided the data used in Figure 10.4.

Figure 10.4 compares the incidence of different clutch sizes in Britain and Europe from the studies in Table 10.1, which are mainly contemporaneous, and shows a significant difference. This difference is almost certainly the result of the depression of clutch sizes in Britain since 1952, as comparison of European records with those

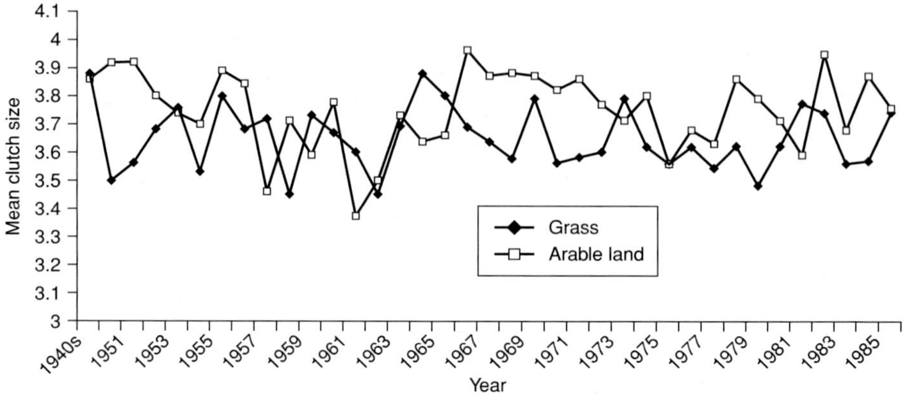

Figure 10.3. Mean annual clutch sizes in Lapwings in England and Wales 1940–1985, in grassland and arable habitats. The 1940s are shown combined as there were only 54 eligible records in total. Source: BTO nest record cards analysed for Shrubb (1990).

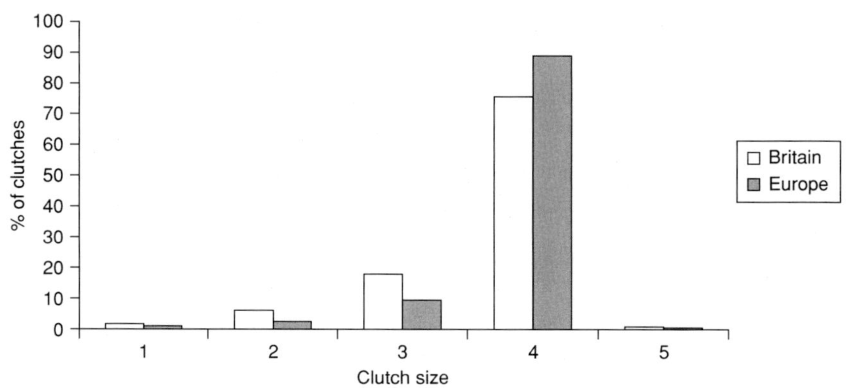

Figure 10.4. Clutch size in Britain and continental Europe compared. There were more small clutches (3 eggs or fewer) in Britain and consequently fewer of 4 eggs ($\chi^2/_4$ = 169.35, $p<0.0001$). See Table 10.1 for data sources.

Table 10.2. Mean clutch size of Lapwings in all farmland habitats in England and Wales before and after 1951. The means are significantly different (Z = 6.98).

Source	*Number of nests*	*Mean clutch size*	*Standard deviation*
Spencer	303	3.83	0.496
Whitaker	125	3.78	0.535
BTO to 1951	120	3.88	0.369
Overall	**548**	**3.83**	**0.465**
BTO 1952–61	1273	3.64	0.665

Sources: Spencer 1953, Arthur Whitaker diaries in Nethersole-Thompson & Nethersole-Thompson 1986 and BTO nest record cards analysed for Shrubb 1990.

up to 1951 in Table 10.2 shows no such significant difference, although marginally fewer clutches of four and more of three were then laid in Britain.

How far clutches of three or fewer represent complete layings is not always clear. At least some are the result of partial loss during laying. Rinkel (1940) noted experimentally that when clutches were removed during laying, leaving at least one, the female completed laying her four eggs and incubated what was then a partial clutch. Under field conditions Galbraith (1988a) found that this particularly occurred in tilled habitats, where females which had begun laying lost one or two eggs to cultivations and completed their now depleted clutch in a fresh nest. In tilled land in England and Wales there were significantly more small clutches on bare ground than in emerged crops, suggesting that the same phenomenon operated there (Shrubb 1990). Similarly I found fewer small clutches in rough grazing than in improved grass, which is often chain harrowed or dressed with farmyard manure in spring. Swindells (1997) cited chain harrowing as an important cause of nest loss (Chapter 11).

Evidence of a significant seasonal decline in clutch size or a decline in clutch size in repeat layings is limited. However Chamberlain & Crick (2003) observed a seasonal decline in clutch size during the decade of the 1980s in England and Wales and Hart *et al.* (2002) noted a similar decline during 1995–97 on the North Kent Marshes. Some studies have found that repeat clutches may be smaller. Thus of 23 females laying four eggs for their first clutch in The Netherlands, three laid only three with their second, giving a mean of 3.87 (Klomp 1951). Jackson & Jackson (1975) found that repeat clutches were consistently smaller in their New Forest study over four years, with an average difference of 13% (mean 3.36 *cf* 3.87 eggs) and Kooiker (1987) noted that all the few clutches of two he found were second clutches. It is not a consistent trend, however, for Baines (1989) found that replacement clutches were significantly smaller in unimproved grassland but significantly larger in improved grassland.

LAPWINGS' EGGS

Lapwings' eggs are pyriform (pear shaped) and unglossed. They are arranged in the nest with points facing inward. Their background colour typically varies through shades of brown, umber, olive, clay or stone colour, blotched, streaked, spotted or dusted with black (Cramp & Simmons 1982). Such coloration continues the theme of cryptic camouflage as a defence against predators (Plates 12 & 13). Cramp & Simmons noted that eggs are occasionally found with background colours of blue, green, red (erythristic) or pink but they are rare. Nethersole-Thompson & Nethersole-Thompson (1986) noted that erythristic eggs were marked with reddish rather than the usual black.

Lapwings lay large eggs. Measurements of 500 eggs across the range gave mean weights of 25–26g (range 22–29g) and mean dimensions of 47×33mm (range

39–53×30–36mm), with no significant geographic variation (Cramp & Simmons 1982), although Liker (1992b) observed that his Hungarian population, in common with other Eastern European populations, laid smaller eggs than Western European Lapwings. Thus the normal clutch of four eggs represents about 45–46% of the female's weight.

Egg weights may vary with habitat. In particular eggs laid by females nesting in arable land have been found to be heavier than those laid by females nesting in grassland (Murton & Westwood 1974, Galbraith 1988b, Blomqvist & Johansson 1995). Murton & Westwood also found that females nesting in fen grassland laid larger eggs than those nesting in heathland. Egg sizes laid by the same female may also vary seasonally, as several studies have found that those in replacement clutches tend to be smaller (Galbraith 1988b, Blomqvist & Johansson 1995, Hegyi & Sasvári 1998). However, this is not a consistent pattern, as Hart *et al.* (2002) found no such difference in the North Kent Marshes and Galbraith noted it only in replacement clutches on his rough grazing site.

Several studies have shown that it is the female's body condition, rather than size, that determines egg volume; heavier females lay larger eggs (e.g. Galbraith 1988b, Blomqvist *et al.* 1997). Although this relationship was not observed by Hegyi & Sasvári (1998), they found that only heavier females laid replacement clutches, again suggesting that lighter females had poorer reproductive capacity. As discussed in Chapter 12, large eggs result in larger chicks at hatching; which survive and grow better than hatchlings from small eggs. Galbraith (1988b) made the point that clutch size in Lapwings is largely fixed (Figure 10.4) and differences in habitat and individual quality are thus more likely to be reflected in egg size and quality rather than in the number of eggs produced.

Ultimately the female's ability to produce a large clutch early in the breeding season depends on the food supply available to her in the pre-laying period. An abundant food supply promotes better body condition (weight) and therefore larger eggs and a better chance of the chicks surviving. The pattern some studies, discussed above, have established, of larger eggs being laid by females in arable habitats, raises the question of whether it is the early food supply which is easily available in arable habitats with newly tilled land that at least partly explains the anomaly of Lapwings persisting in nesting there despite declining breeding success (Chapter 4). They are not only drawn by safer nest sites on tillage but also by a favourable and easily obtained early food supply. This idea has not been tested and it seems desirable to do so. Mark Bolton has suggested to me that an alternative explanation may be that optimum egg size differs between arable and grass habitats and females need to lay larger eggs in arable land as chicks need greater yolk reserves in the period immediately after hatching, when they have to move, sometimes substantial distances, to foraging areas. But, if the food supply explanation is correct, some spring crops may now have become a serious ecological trap because modern management changes have altered the habitat resources for the later stages of the breeding cycle to the Lapwing's disadvantage, as noted by Galbraith (1988b).

INCUBATION

Incubation in Lapwings starts when or just before the last egg is laid, although eggs may be covered in some circumstances, such as bad weather, before incubation proper begins. Both sexes incubate but the female takes the larger share by day and, according to Nethersole-Thompson & Nethersole-Thompson (1986), only the female incubates at night. Spencer (1953) noted that females were on the nest on 84% of visits he made over four years, males on 16%, but males have been recorded on the nest as infrequently as on 1% of visits and as incubating for as much as 50% of nest watches (Wickens 1948, Tully 1948). Spells taken by the male and their frequency vary but I agree with the Nethersole-Thompsons that he rarely sits for more than *c.*1.5 hours and often for no more than ten minutes or so at a stretch. Nor does he necessarily stay on the eggs until the female returns. In polygynous pairings I have watched the males relieved all their females but this does not always occur. Liker & Székely (1999b) noted that only seven out of 12 polygynous males in their study incubated the clutches of both their primary and secondary females and Wilson (1967), watching a trigamous pairing in Lancashire, never observed the male to relieve any of his mates. Nevertheless Liker & Székely (1999b) found that the time spent incubating by monogamous and polygynous males did not differ. They also observed that, although all the 29 monogamous males they watched remained on their territories until eggs hatched or failed, four of the females deserted between two and ten days before hatching, leaving the male to hatch and rear the young.

There is no special change-over ritual, although copulation often takes place then. The female may simply slip off and go to feed or drink, with the male flying across to take over after an interval or apparently when he notices her absence. Or there may be a straightforward take-over, one in and one out. Or sometimes the male will initiate a change-over, walking to and round the incubating female and clearly inviting her to move over. The Nethersole-Thompsons also record the male occasionally calling the female off and escorting her to the feeding grounds, subsequently escorting her back almost to the nest. Females may spend part of the time off the nest bathing and preening and one on my Welsh site used pools of seawater on the shore nearby. The relieved female may walk away to feed, sometimes tossing grass over her shoulder as she does so, or fly to neutral ground. On returning she lands a little way from the nest and the male, if on, leaves, the female then trotting or walking on.

Incubating females can be surprisingly restless, getting up and turning round, resettling the eggs, adding to the nest and so on. They often reach out and pluck vegetation from around the nest to add to the lining but for much of the time they sit right down in the nest so that only their back is visible, apparently dozing. In this posture they are strikingly inconspicuous against their preferred backgrounds and their presence is most likely to be revealed by small head movements. They are, however, alert to danger. Iversen (1986) measured the time spent incubating and off the nest at seven nests in Danish saltmarsh. He found that, overall, birds incubated for 77% of the observation period (146 hours), were off without disturbance, to feed etc., for 13% of the time and were off the nest due to disturbance by predators or humans for 10% of the period. Interestingly, human disturbance caused most disruption on this site, accounting for 80% of the time off the nest due to disturbance. Iversen also observed that birds sat tighter as incubation progressed, with the time spent off the nest to feed etc. declining by some 70% after the first nine days of incubation.

Incubation periods vary but ten European studies gave a mean of 26–27 days, whilst individual periods varied from 21 to 34 days (Table 10.3). The New Forest sample, from Jackson & Jackson (1975), seems exceptional, lying well below the mean recorded in the other studies. Interestingly this was a heathland site and soil temperature may have influenced this, as heathland soils warm quickly, perhaps aiding incubation. Galbraith (1988a) found that incubation periods declined over the laying season, suggesting that this arose because the birds intensified incubation effort with later clutches. Nethersole-Thompson & Nethersole-Thompson (1986) observed that longer incubation periods were partly accounted for by light sitting and frequent disturbance, particularly by sheep, keeping birds off the eggs during the day. Embryos are exceptionally resistant to cold weather (Heim 1974) and eggs may still hatch even if abandoned for 48 hours (Laven 1941).

Table 10.3. Incubation periods of the Lapwing observed in European studies.

Place	*Number of nests*	*Incubation Period (days)*	*Mean (days)*	*References*
Europe overall			27	Visser & Beintema 1991
England: New Forest	96		21 ± 1.77SE	Jackson & Jackson 1975
England: North	11		27.5	Spencer 1953
England: North	4	24	24	Brown 1926
Scotland: Central	15	21–28	25.2	Galbraith 1988b
Scotland: North	41	24.7–34	28.1	Nethersole-Thompson & Nethersole-Thompson 1986
Wales	15	26–29	27.4	Thomas 1938,1939
Switzerland	22	26–29	26.8	Heim 1974
The Netherlands	6		27.1	Klomp 1951
The Netherlands			28	Beintema & Visser 1989a.
Finland			24	Vepsäläinen 1968

HATCHING

Hatching is a lengthy and exhausting process. Chicks start calling in the egg several days before actually emerging and the parents call to them. They also call to them on returning to the nest, using the 'all clear' call, a soft 'pee-wi', (Spencer 1953). The incubating female's behaviour also changes. She is less relaxed, more fidgety, and sometimes makes what appear to be preening movements but which presumably check what is happening or are a reaction to calls from the chicks. Nethersole-Thompson & Nethersole-Thompson (1986) noted that, when a chick is emerging, the brooding adult, usually the female, stands up and calls to it. The brooding adult also fluffs itself up to accommodate and dry the emerging chicks (Plate 11). The cock will also sit close by the nest during hatching and will brood the chicks, once dried off under the hen, as others emerge.

Hatching may be achieved in a period of hours or it may take several days, depending partly on how it is reckoned. Jackson & Jackson (1975) gave 9–26 hours for a full clutch of four eggs from the time the first chick emerged. Nethersole-Thompson (in Bannerman 1961) noted that all the chicks in a brood normally hatched within seven to twelve hours but twice noted spreads of 24 hours. I have few personal records but these also include one record of a brood taking a full 24 hours to emerge.

More frequently authors have calculated the hatching period from the time the first eggs start to chip. On this basis Brown (1946) estimated hatching periods for three clutches which extended to about 56, 58 and 71 hours respectively. Nethersole-Thompson in Bannerman (1961) gave 36 ± 2 hours to 96 ± 9 hours for 31 eggs, 18 of which hatched in 49–60 hours. In Switzerland Heim (1974) measured 55 chipping periods which gave 26 periods of two days to hatching, 16 of three, 11 of four and two of five, for a mean of 2.8 days. No significant variation was apparent between chipping periods in April and May/June. Empty shells are often removed and dropped some distance from the nest. The adults also characteristically bury small pieces of shell below the nest lining and their presence is a good guide as to whether the nest was successful or not.

CHAPTER ELEVEN

The breeding season: nesting success

Studies in Britain and Europe over the past 30 years or so have shown that the proportion of Lapwing females in any population or group succeeding in hatching a clutch ranged between 35% and 73%. Success rates varied for a number of reasons but, as Beintema & Muskens (1987) pointed out, in evaluating the impact of nest loss on productivity in Lapwings, the ability to replace clutches is a crucial factor.

NEST SUCCESS

In Lapwings there are marked differences in nest success with habitat, as noted in previous chapters. For example, studies in Britain have found that nests on tilled land are often more successful than those in grassland, mainly because predation is less in the former and nests are not at risk to livestock (e.g. Baines 1989, Shrubb 1990, Chamberlain & Crick 2003), but this is not invariably so. Galbraith (1988a) and Blomqvist & Johansson (1995) found no difference between their arable and pasture sites. In grassland nests are more successful in unimproved than improved habitats, as shown very clearly by Baines 1989 (Figure 11.1). Shrubb (1990) also suggested that nests in habitats outside farmland were more successful than farmland nests because the former were not subject to the risks imposed by cultivations or livestock. Everywhere studies have found that nests in tilled land are much less successful on bare land than in emerged crops because the former are at risk to cultivations, which have ended in the latter. Thus individual studies show pairs nesting in tillage losing a high proportion of nests on bare land but then replacing them with a high proportion of successful repeat layings (Figure 11.2). Patterns of nest success in tilled land in Europe may not be quite the same as in Britain because of differences in the timing of cultivations, for example in maize, a more widely grown crop in Europe, compared to cereals. Cultivations are also later in the north and may be later in relation to the laying patterns of Lapwings.

Shrubb (1990) found that the percentage of nests lost to agricultural operations in tilled land increased over 1962–85, which appeared to reflect changing patterns of March rainfall and their impact on the timing of farmwork (p.138). Such nest losses may also have increased over a longer timescale with the increasing size and complexity of farm machinery. Lloyd (1912) noted that horsemen commonly shifted nests out of the way when working land, passing over the nest and then replacing the eggs. As I can also attest from my own experience, the adults readily returned to them and made a new nest scrape to hold them.

Both Beintema & Müskens (1987) and Chamberlain & Crick (2003) show a curvilinear relationship between the probability of a nest surviving and time, with early and late nests having poorer survival rates. With early nests this may arise because they are particularly vulnerable to adverse weather, such as spring snow

Figure 11.1. The percentage of Lapwing eggs hatching in different grassland habitats in northern England. Data from Baines 1989.

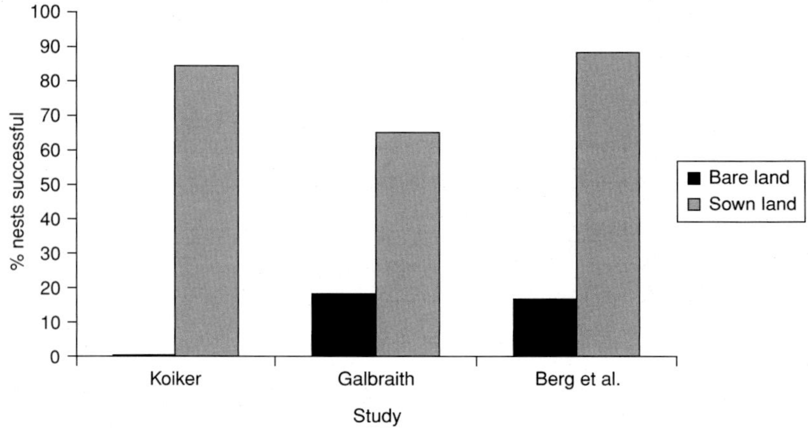

Figure 11.2. The percentage of successful tillage nests of Lapwing on bare or unsown land and of repeat nests on sown land in three European studies. Data from Kooiker (1984), Galbraith (1988a) and Berg *et al.* (1992).

storms. Later nests may be more vulnerable because the number of adults present declines as pairs complete nesting, so that the benefits of group defence decline (Dyrcz & Witowski 1987). Elliot (1985a) also noted an increase in the intensity of anti-predator responses as incubation progressed and the value of the clutch increased. Beintema & Müskens also found that nests were more vulnerable to predation during laying than during incubation and this showed the same curvilinear relationship with time.

The BTO nest record cards I examined for the period 1962–1985 also suggested that hatching success may be poorer with small clutches (three eggs or fewer). Clutches of one and two eggs were particularly prone to desertion and it seems likely that many such clutches, if not already deserted, had been depleted when found and so were more prone to desertion. However, clutches of three were still less successful than those of four and five eggs (Figure 11.3).

Social breeding enhances nest success and so does the openness of breeding sites. Finally Beintema & Müskens (1987) found a relationship between vole abundance and wader nest predation in Dutch grasslands, predation of nests increasing when vole numbers were low as predators, perhaps particularly mustelids, switched to birds' nests; Lapwings and Black-tailed Godwits were the species most affected.

PARTIAL EGG LOSSES

Besides the loss of complete clutches Lapwings lose a drizzle of eggs from otherwise successful nests. This seems not to have been specifically examined in many studies but, in England and Wales up to 1985, the BTO's nest record cards indicated that

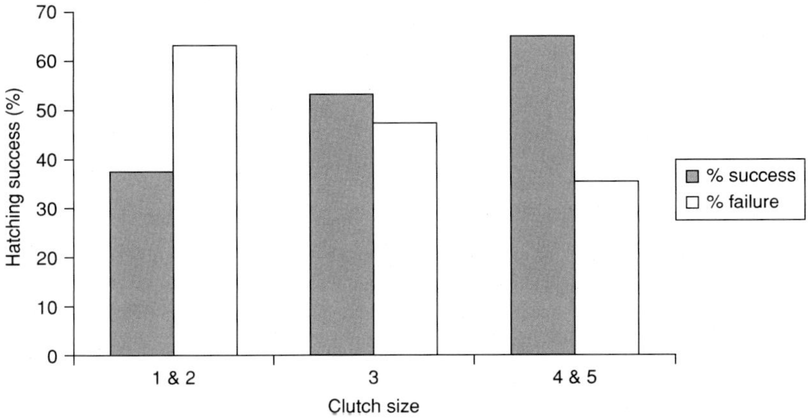

Figure 11.3. The percentage of Lapwings' nests with different clutch sizes which succeeded or failed in England and Wales during 1962–1985. Source: BTO nest record cards analysed for Shrubb 1990. The proportions of successful nests differed significantly ($\chi^2/_2 = 65.38, p<0.001$).

such losses involved *c.*8% of the total of eggs laid in nests for which a successful result was recorded. More partial losses were reported in grassland than tilled land (Shrubb 1990). Where the causes of such failure was known infertility or the death of embryos were the most frequent. These were rarely noted as causes of total clutch loss, perhaps because the eggs quickly disappear. The pattern of occurrence of infertile eggs and dead embryos lends support to the idea that organochlorine pesticides influenced breeding performance in the 1950s and 1960s, discussed on p.140. The percentage of eggs lost to these causes rose very sharply after 1950 and it remained higher than before that date into the 1980s (Figure 11.4).

CLUTCH REPLACEMENT

Although Klomp (1951) showed experimentally that Lapwings could be induced to lay up to five clutches in a season, this is not typical of modern farmland conditions at least. Not all females that lose their first clutch relay. In The Netherlands Beintema and Müskens (1987) noted that all relay in the first weeks of the laying season but, as the season progresses, the probability of relaying declines, at least partly because females that start laying later in the season have less time for repeat clutches before moult, when laying ceases. Matter (1982) noted the same pattern in Switzerland and also recorded that replacement clutches, if lost, were not replaced at the same rate as first clutches. Jackson & Jackson (1975) found that 83% of first clutches lost over three years in their rough grazing and heathland site in the New Forest were replaced.

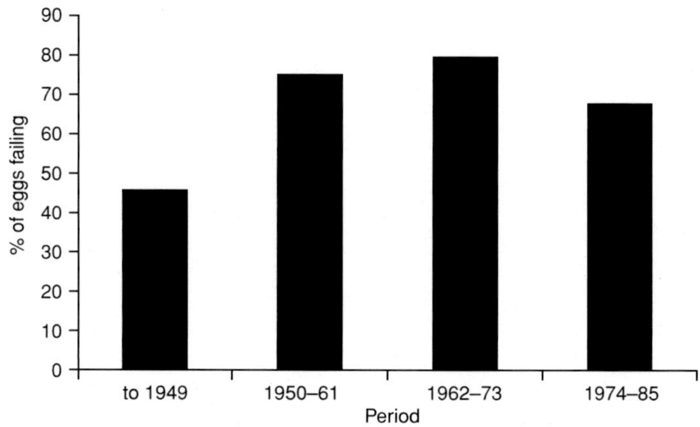

Figure 11.4. The percentage of Lapwing eggs failing due to infertility or the death of embryos in otherwise successful nests in England and Wales in four different periods. Percentages are of all eggs failing in such nests for which a definite cause was recorded. Source: BTO nest record cards analysed for Shrubb 1990.

In improved farmland various studies have shown that the overall rate of clutch replacement today is considerably lower than indicated above, because rapid and earlier crop growth and, in grassland, greatly increased stocking rates, curtail laying (see below). The importance of crop growth in curtailing the laying season was neatly underlined by Galbraith (1988a). On his arable site all the clutches in mid- to late May were laid on fields cultivated six weeks later than normal, when crop growth had rendered all other cereal fields unsuitable. At least two of these clutches were laid by females that had failed in more typical cereal fields 3–4 weeks earlier and had temporarily joined flocks of failed breeders. Parish *et al.* (1997a) noted that in upland grassland only 61% of females which lost their first clutch relaid, whilst Baines (1989) observed that of 254 first clutches in grassland which failed, only 125 (49%) were replaced, with marked variations in the rate of replacement with habitat (Figure 11.5). The rate of replacement in arable land was similar to that observed by Berg *et al.* (1992) in their largely arable study area in Sweden (66%). Their area was mainly devoted to cereals but this is certainly a lower rate of replacement than in the past, when summer fallows were extensively available and laying could continue later as the birds were always able to nest on bare ground (p.138).

The ability to replace failed clutches is an important component of productivity in Lapwings and its restriction in modern farmland an important factor in the species' decline there. That the underlying cause is change in agricultural management is clearly indicated in Figure 11.5. Lapwings nesting in improved pasture and meadow (grass harvested for forage) had replacement rates less than half of those of birds nesting in unimproved pasture. Laying can also continue longer in habitats outside farmland (Figure 10.2). A good example was provided by

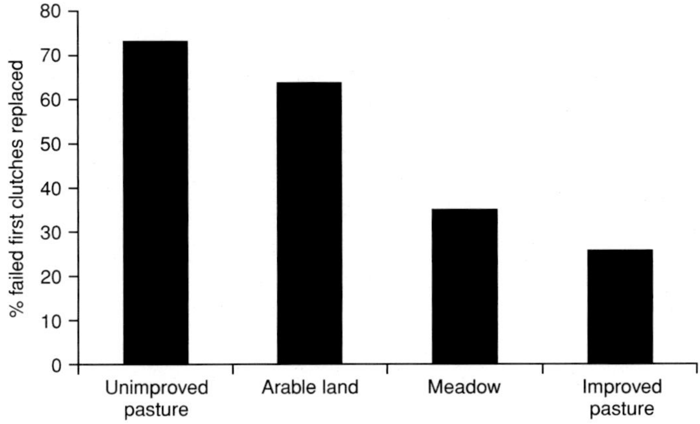

Figure 11.5. Replacement rates of failed first clutches by Lapwings in different habitats in northern England. Data from Baines 1989.

Iversen (1986), who found that females on saltmarsh had an average of 2.4 nests per season, a higher rate than any farmland study I have examined.

The time Lapwings take to replace a clutch varies. Nethersole-Thompson & Nethersole-Thompson (1986) recorded that the first egg of the replacement clutch was laid five to seven days after the first clutch was lost but Hegyi & Sasvári (1998) found that the shortest period was seven days and the longest 17, whilst Rinkel (1940) gave an interval of seven to eight days and Berg *et al.* (1992) noted a mean of nine days. Klomp (1951) gave an average of 12 days to produce a new clutch and I have recorded intervals of 10–14 days from loss to the start of incubating the new clutch. Both Klomp (1951) and Parish *et al.* (1997a) noted that a new clutch can be produced after the early loss of a brood but the loss of chicks usually terminates breeding. However Hegyi & Sasvári (1998), who noted that only 59% of females relaid in their grassland area in Hungary, found that there was a significant tendency for them not to relay if nests were lost after 15 days of incubation.

Although replacement clutches are most usually laid in the same territory, females may switch territories for later clutches. Spencer (1953) gave an interesting case of a female, whose eggs were identifiable, that lost two clutches in her original territory, shifted to a completely new area for her third attempt, lost that and returned to her original district for her fourth and finally successful clutch.

CAUSES OF NEST LOSS

Lapwings lose nests for variety of reasons but the principal causes are predation, cultivations, particularly in tilled land, trampling by livestock and desertion (Table 11.1). Desertion may be the result of factors such as spring snow storms or flooding,

Table 11.1. Causes of nest or egg loss in Lapwings recorded in individual European studies.

Place	*% of nests/eggs lost to*						*Reference*
	Predation	*Farm work*	*Trampling*	*Desertion*	*Other*	*Cause unknown*	
England							
New Forest	*26.8*		*1*	*2.8*	*1.4*	*1.6*	Jackson & Jackson 1980
Kent[1]	90		7		2		Hart *et al.* 2002
N. England							Baines 1990
Unimproved grassland	48	2.6	2.7				
Improved grassland	65.8	12.4					
Arable land	24.6	11.5					
Scotland[1]							Galbraith 1988a
Arable land	*59.3*	*29.7*			*3.4*[2]		
Rough grazing	*89.4*				*5.2*[2]		
France							Trolliet 2003
Vendée	20.8						
Arable land		17.5					
Grassland		9.0	4.3				
Alsace arable land	31.0	29.1					
The Netherlands[1]	44	7.1	22.7	7.4		18.9	Beintema & Müskens 1987
Denmark							
Tipperne NR	28.8		?	0.5	1.4		Thorup 1998
Sweden							
Arable/grassland	7.1	43.5			0.6		Berg *et al.* 1992
Grass	59.5						Ottvall 2005
Coastal mixed farmland	16	6	1	3	4		Blomqvist & Johansson 1995
N. Germany							
Arable/grassland	26	14	2–9	1		10	Matter 1982
Switzerland							
Arable land	29	14		2	3	9	Matter 1982
	27.6[3]			*11.6*	*2.4*[4]		Heim 1978
Hungary							
Grass	21.7		5	1.7	5	13.3	Liker 1992a,b

Percentages are of the total number of nests studied except where indicated. Figures in italics are percentages of eggs lost. Farm work includes cultivations and other activities.

Notes:

In Scotland a further 6.5% of eggs were recorded as infertile or having dead embryos.

[1] Percentages are of nests lost not total nests studied.

[2] Includes nests trampled by livestock.

[3] Percentage is of eggs lost to predation and farm work combined.

[4] Infertile eggs.

? Thorup (1998) expected 5–15% of nests to be trampled.

themselves important secondary causes of loss, but it is often livestock related (see below) and, in Britain at least, occurs more frequently in grassland habitats (Shrubb 1990, Chamberlain & Crick 2003). Chamberlain & Crick also observed that nest predation increased during the 1990s, as did Teunissen *et al.* (2005) in The Netherlands. Trampling by livestock does not emerge from the table as an important cause of nest loss in many areas but this may simply arise because, on many of these study sites, livestock numbers or management were not important or the information was not fully recorded, e.g. at Tipperne. In the general study of The Netherlands a very different picture emerged, although the percentages are not strictly comparable because of differences in the way they are calculated. In the BTO nest record cards, only a proportion of which record causes of loss, trampling by livestock accounted for 16% of clutches for which the cause of loss was known in grassland in England and Wales up to 1985 (Shrubb 1990), a not dissimilar result to that in The Netherlands, where the percentage was similarly calculated. On the same basis desertion accounted for 22% of losses in grassland in England and Wales in the same period but, for all nests for which a result was known, losses were 8% to trampling and 10% to desertion.

In Lapwings there is an important difference in the nature of predation as a cause of nest or egg loss and agricultural operations. Predation has been shown to be density dependent, i.e. it fluctuates with numbers, but it is often negatively so, as social breeding reduces predation (see below). The impact of agricultural operations, however, is independent of the density or numbers in the nesting group at risk. Ploughing a field will destroy all nests irrespective of whether one or fifty pairs are nesting there. Unlike those to predation, losses to cultivation today also tend to be strongly clumped into the early part of the nesting season (Figure 11.2), although this does depend upon what crops are grown.

The relationship with livestock is different again and depends on the density and types of grazing animals and the length of time nests are exposed to them. This is considered in detail in the next section but the main cause of loss to livestock is to trampling although, particularly with sheep, desertion becomes a problem at high stocking densities.

PROBLEMS WITH LIVESTOCK

Ground-nesting birds in agricultural grasslands have always had to cope with the presence of livestock which, in modern farmland where they occur at high densities, have become a significant problem. Lapwings incubating in grassland react differently to different types of livestock in their nesting fields. The effect of livestock on incubating behaviour also depends on stocking densities, on the age of livestock—young cattle create more disturbance than adults—or on whether fields are stocked with a set number of animals, paddock grazed or ranched i.e. where animals are allowed to roam over several fields at will. Paddock grazing involves regular shifts of stock which may be particularly disruptive. Stocking densities (animals/ha) are more significant than stocking rates (livestock units/ha) because a given level of livestock units gives rise to very different densities. A dairy cow is normally rated at one livestock unit but other cattle vary from 0.4 (animals under one year old) to 0.8 (beef cows) and ewes are 0.15, the different ratings reflecting food intake (Coppock 1976).

Beintema & Müskens (1987) showed differing rates of nest survival with different patterns of stocking common in The Netherlands, illustrating clearly the rapid decline in nest survival with rising stock densities. In terms of stocking rates they showed sheep and young cattle to be more damaging because of the difference in densities at the same stocking rates as adult cows.

In many parts of Britain which are devoted to rearing livestock for meat, experience seems to differ and trampling by livestock seems a less significant cause of nest loss, despite high stocking rates. M. O'Brien (in Taylor & Grant 2004), studying a range of upland sites across northern Britain, found that nest losses through trampling by sheep tended to be low on upland grasslands. Similarly Baines (1990) found that direct effects through trampling on his upland sites were low but suggested that the high stocking densities on his improved grassland sites there resulted in low rates of clutch replacement. Experience in Wales, where stocking rates of sheep are particularly high, supports this suggestion strongly. Nevertheless an increasing trend in nest losses to trampling recorded by the BTO nest record cards for 1962–1985 correlated significantly with both rising cattle and sheep numbers (Shrubb 1990). Sheep at high densities are also likely to cause Lapwings to desert sites (see below). Chamberlain & Crick (2003), for the whole of the UK during the 1990s (the only decade for which the information was available) also observed that the presence of grazing animals in nesting fields in grassland resulted in later laying, smaller clutches and higher clutch failure rates, as did Hart *et al.* (2002).

I found that three causes of nest loss in grassland were significantly correlated with rising stock densities, losses to farm work, which increases with stocking levels, desertion, and trampling by livestock (Shrubb 1990). These causes of nest loss thus appear to be particularly related to increasing intensification in farm management in grassland. Figure 11.6 shows the percentage of grassland nests recorded as lost to these causes by the nest record cards at intervals from 1945 to 1985 in relation to numbers of cattle and sheep per 1,000ha of grass of all types, which provides a simple index of management intensity. It shows an interesting pattern, rising steeply until stock densities reached *c.*4,500 animals/1,000ha and then declining. As the rise in stock densities was virtually continuous, this is an expression of change over time as well and so years are also plotted.

Stock densities eventually stabilised at around 6,000 animals/1,000ha in the 1990s, when Chamberlain & Crick (2003) showed, for the whole UK, that these losses (which they classified as 'destroyed') had declined, having peaked in the 1980s. In the absence of any significant decline in stocking densities (June Census Statistics) there seems little reason to doubt that the anomalous pattern shown in

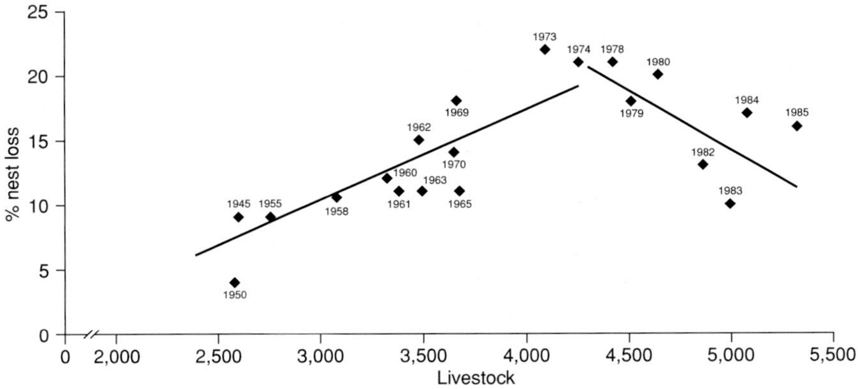

Figure 11.6. Number of Lapwings' nests in grassland in England and Wales lost to desertion, trampling and farm work as a percentage of all grassland nests for which the cause of loss was known during 1945–1985 in relation to total animals per 1,000ha of all grass. Source BTO nest record cards analysed for Shrubb 1990.

Figure 11.6 has arisen because Lapwings have increasingly ceased to attempt breeding in grassland areas subject to high stocking densities. Clear support for this idea was provided by the 1998 Lapwing survey, (Chapter 4 and below p.161).

With cattle my own experience is with rather low stocking densities of one or two animals per hectare, and incubating Lapwings can be remarkably indifferent to their presence in these circumstances. One group watched in Wales recently settled in an 18ha field where 17 cows and four young followers were out-wintered and remained through the breeding season. In watches on 57 days over two breeding seasons, I saw no evidence that these Lapwings took any notice of the grazing cattle, which did not disrupt incubation or cause females to leave the nest at any time I was watching. Cows grazing near a sitting bird elicited little reaction, although the sitting birds clearly watched them, sometimes cocking their heads to one side to peer up at them. One female Lapwing I was watching feeding ran rapidly back on to her eggs when she noticed a cow grazing steadily towards the nest, whilst another trotted about in front of and beside a grazing cow approaching her nest and appeared to divert it. I often wondered, in fact, if the cattle were aware of the birds and avoided them and, at least once, observed a cow change direction and walk round a sitting bird. Similarly I never observed Lapwings to be kept from their nests by the presence of our cattle on the family farm in the past, where our stocking densities were similarly rather low (see above). My experience suggests that this kind of relationship between nesting Lapwings and cattle was the norm before the massive increase in stocking rates and densities that has occurred since the early 1960s (Figure 4.4). It also seems to exist in Scandinavia in areas with similar low stocking densities, for example Ettrup & Bak (1985) in Denmark and Ottvall (2005) in Sweden.

Young cattle, which are intensively inquisitive, pose more problems as Beintema & Müskens (1987) showed. I have watched incubating Lapwings drive six-month-old

Figure 11.7. Defence of nest or young. (i) adults labouring up to see off high-hovering Kestrel, often unsuccessful in preventing chicks being taken. (ii) adults circling and calling over marauding Fox. (iii) adult driving off calf from the vicinity of nest in banner display.

calves away from apparently investigating their nests, standing up suddenly in front of their noses and opening their wings, flashing their wing linings at them (Figure 11.7). This was usually effective in diverting such activity from the nest and Nethersole-Thompson (in Bannerman 1961) also noted that the sitting bird may run at an animal with its wings spread and calling. This display seems to be a general nest defence mechanism amongst lapwing species against large herbivores. Hayman *et al.* (1986) noted that similar displays by White-headed Lapwings were normally sufficient to prevent nests being trampled by hippos and buffaloes, E. Andrews (pers. comm.) has seen Blacksmith and Spur-winged Lapwings divert elephants in the same way and del Hoyo *et al.* (1996) provided a splendid photograph of the former species inviting a Warthog to depart. Nevertheless Beintema & Müskens (1987), in their studies in Dutch grasslands, noted that waders generally did little to defend their nests against livestock, often flying off at the last minute to avoid being stepped on. As Lapwings have well developed defence mechanisms against such events, this rather suggests that the high stocking densities practised swamped their reactions. Nevertheless their figures show that Lapwings lost many fewer nests to trampling by any livestock than the other waders they studied, except Oystercatchers, which suggests that some form of fairly effective nest defence was involved. Thorup (1998) made the same point for waders at Tipperne.

My observations differed sharply from those of Hart *et al.* (2002) in the North Kent Marshes, who found, as did Chamberlain & Crick (2003) from the BTO nest records, that the presence of livestock, even at the low stocking rates practised of 0.2–0.51 livestock units/ha, resulted in Lapwings laying later, laying smaller clutches and losing more nests in grazed than ungrazed marshes. The main cause of loss was considered to be predation, not trampling by livestock, because the presence of livestock disrupted incubation, increasing the risk of predation whilst birds were off the nest. However, it may sometimes be difficult to distinguish between trampled nests and those lost to predators. Recent work by the RSPB with cameras placed at nests has recorded several instances of nests being trampled and then scavenged by a mammalian predator attracted by the scent of the broken eggs (M. Bolton pers. comm.)

Two difficulties arise in comparing the findings of Hart *et al.* (2002) with my own experience. One was that the stocking pattern practised was often ranching, allowing stock free range over several marshes. In such systems stocking rates in individual plots can vary with the grazing patterns of the animals and temporarily may often considerably exceed the overall average. Secondly their examination of the impact of grazing stock on incubation patterns was only made with sheep.

A number of studies besides Hart *et al.* have observed that sheep disrupt the incubation behaviour of Lapwings, keeping birds off nests, and sometimes out of the nesting field, for long periods and, if stocking densities are very high, as they often are in Wales, causing site desertion (e.g. Nethersole-Thompson & Nethersole-Thompson 1986, Shrubb *et al.* 1997, Taylor & Grant 2004, pers. obs.). I am not aware that any study has answered the question why. Beintema & Müskens (1987) suggested that individually sheep did little harm; it was the stocking densities that did the damage. At any given stocking rate this will be up to ten times higher than with dairy cows. There are indications in the literature that sheep are more difficult for Lapwings to repel from nests, as they seem to disregard threat displays, something which I have also observed. Thus both Took (1936) and Nethersole-Thompson (1940) described incidents where vigorous displays followed by physical assaults failed to drive sheep away from nests. Sheep may also act as predators of nests of ground-nesting birds (Pennington 1992) and the work with nest cameras by the RSPB has now recorded several instances of sheep eating Lapwing clutches (M. Bolton pers. comm.). Another problem might be that the short swards produced by intensive grazing by sheep attracts crows (Fuller & Gough 1996), which are important predators of Lapwings' eggs, creating a toxic alliance of regular disturbance and attracted predators. Continual disturbance also lengthens the incubation period, so that nests are exposed to predation risk longer. In Western Europe these also appear to be problems which occur particularly in the UK, which holds half the EU sheep flock (Eurostat 1995). The difference in flock sizes in the UK and other EU countries is very marked (Figure 11.8).

NEST DEFENCE AND PREDATORS

In Britain examination of the 19th century county and regional avifaunas makes quite clear that, for the period for which we have reasonable ornithological records, Lapwing populations and populations of their predators were both at their most abundant at the beginning of the 19th century. Predation has always existed and large predator populations do not necessarily result in small populations of species

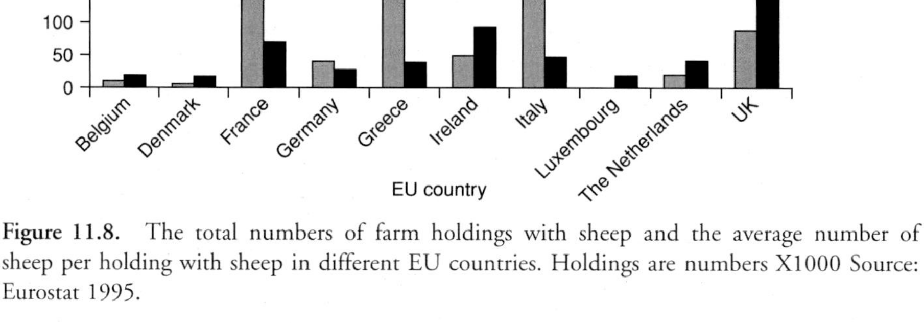

Figure 11.8. The total numbers of farm holdings with sheep and the average number of sheep per holding with sheep in different EU countries. Holdings are numbers X1000 Source: Eurostat 1995.

such as Lapwings. All raptor and corvid populations were severely reduced by game preserving interests throughout the 19th and early 20th centuries and raptors were further reduced, in some areas to the point of extinction, by the side effects of organochlorine pesticides in the 1950s and 1960s. The county avifaunas provide little evidence that these changes resulted in extensive increases in prey species. Species such as Red Grouse, the preservation of which was a prime reason for exterminating predators in the uplands, in fact declined for much of the 20th century (Brown & Grice 2005). We do not know what effect predator control had on Lapwings, as that species also declined with severe overexploitation in the 19th century (Chapter 3).

The modern conservation ethic and the decline of keepered estates has led to recent major increases in predators of Lapwing nests and chicks although, historically, these increases are, in great part, more accurately considered as population recoveries, from the impact of the organochlorine pesticides as well as from declining persecution. Large scale habitat changes have affected this. For example foxes were once scarce in the hills in parts of central Wales but afforestation, and the temporary surge of vole populations it triggered, encouraged foxes to colonise and they then found sufficient food scavenging around increasing hill sheep flocks to maintain themselves after the vole flush was past (Lloyd 1980). As noted below, increases in predator populations interact with modern farmland management in their effects on Lapwings. In grassland, improvement by drainage and reseeding has also severely fragmented most of the favourable areas of unimproved pasture. Such fragmentation reduces Lapwing population densities and group sizes and therefore the effectiveness of nest defence (see below).

The Lapwing has three defences against nest predation. The first is by habitat selection, which involves choosing nesting sites which provide cryptic and broken backgrounds to help conceal the sitting bird and siting the nest away from features such as trees, which harbour predators (Chapter 10). One of the problems faced today by Lapwings nesting in grassland is that modern management has seriously eroded many natural nest defences by drainage, reseeding and increased rates of fertiliser application earlier in the season, to support higher stocking densities, particularly of sheep. The resulting bright green and very uniform swards offer poor security to nests. This was discussed more fully in Chapter 4 and the impact of such grassland improvement on nesting success in Lapwings was set out very clearly by Baines (1989, 1990). In Britain the impact of these changes has been most marked in the standard statistical regions of southwest England, the West Midlands and Wales (see map in Wilson *et al.* 2001) where an expanding pastoral monoculture based on sheep has emerged in many districts. Here 26% of the agricultural area of Britain holds almost half the farm holdings with sheep and 45% of all sheep (June Census Statistics 1997). Lapwings in these regions have virtually ceased to breed in improved grassland. For example only 40 pairs were found on the 60,000ha of improved grassland surveyed there for the 1998 Lapwing survey.

Defence of the nest is by both sexes, with the female coming straight off the nest to assist the male if he has difficulty in driving predators away from the nest area or if he is slow to react. The predators are usually crows, which often work in concert. Elliot (1985b), in his experiments with dummy crows, found that females initiated the response in half the trials when both adults were present, males in 40% and both in 10%. Nesting in groups increases the effectiveness of such aerial defence and active groups of nesting Lapwings are well able to exclude crows from nesting areas or greatly depress their activity (Elliot 1985a, Seymour *et al.* 2003, pers. obs.). Elliot noted this effect even if crows had used the nesting areas before the Lapwings settled. Nevertheless my experience suggests that the effectiveness of such exclusion may partly depend on whether or not crows use nesting fields extensively for normal feeding. This could partly explain the pattern of lower predation rates of Lapwing nests on sites in arable land and the higher rates on improved pasture than on unimproved recorded by a number of studies, which broadly reflects the feeding preferences of corvids in farmland (e.g. Shrubb 2003; Barnett *et al.* 2004). Crows seem more likely to learn that Lapwings are nesting in a field if they feed there regularly. On Elliot's study area (Elliot 1985a,b) site use by crows increased, partly attracted by supplementary sheep feed put out on the grass after the Lapwings settled, a frequent problem, which attracts Ravens as well as crows. They could not prevent loss of nests to crow predation but significantly reduced the risk, repelling all 101 attacks he noted in 109 hours observation. The intensity of mobbing attacks increased as incubation and the breeding season progressed, reflecting the increasing reproductive value of the clutch.

Nesting Lapwings mob crows closely and severely, diving at them from the front and behind and harassing them right out of the territory. I have not seen them actually strike crows but Elliot (1985b) found that the dummy crows used in his experiments were often struck, sometimes with considerable force and usually with the feet. Lapwings can also clearly distinguish between potential marauders and corvids simply flying over the nesting field to feed elsewhere, which elicit no response. This may itself be a defensive mechanism, not drawing attention to their presence. Fences crossing nesting fields may be a particular problem, as crows take cover from aerial attack close behind their base. It is always pleasing to see one promptly set upon by a pair nesting on the other side. Elliot (1985a,b) found that the zone protected by active defence around the nest extended 30–50m from it, increasing to 40–60m in larger groups. There is also a strong negative correlation between Lapwing colony size (the number of nests within 100m) and predation risk, larger colonies being less affected by predators (Elliot 1985a, Berg *et al.* 1992, Berg 1996, Seymour *et al.* 2003). Such negative density dependence may affect population stability, as predation will increasingly affect productivity as population density declines or as local breeding habitats are fragmented (Seymour *et al.* 2003).

Carrion Crows are not, of course, the only avian predators of Lapwings' nests. All corvids take eggs and nestlings and Ravens may become a particular problem being more formidable and dangerous than other corvids. Dick Squires (pers. comm.) commented that the breeding success on one of his sites at Ynys-hir in 2005 was affected by a single Raven that specialised on Lapwings' eggs. Gulls, particularly Black-headed Gulls and Common Gulls, are significant nest predators in some areas and the presence in a nesting field of species such as Grey Herons, which prey on chicks, will provoke mass mobbing attacks. So will cock Pheasants. However, the Lapwing's defence mechanisms are the same for these different species.

Except for harriers, raptors are probably not significant predators of Lapwings' nests but both Common Buzzards and Common Kestrels can be significant predators of their chicks. Buzzards are severely mobbed on the ground and in the air, usually by a high proportion of the birds present. The Kestrels' technique for hunting chicks, high circling round the breeding site and an ability to detect chicks at long distances, is difficult for adults to counter (Shrubb 1993. Figure 11.7). Although wader and tern colonies and game-rearing fields can be ravaged by them, however, wader and other chicks are a second-best option for Kestrels and tend to be taken mainly when voles are in short supply. Work with captive Kestrels by J.G. Kirkwood (quoted by Cramp & Simmons 1979) suggested that day-old chickens weighing 36g had about half the nutritional value of laboratory mice weighing 19g. In the wild, small rodents are also likely to be more worthwhile prey than nestlings. Other falcons are, perhaps, fairly infrequent predators of nesting Lapwings but one passage Merlin I watched sitting on a fence near three nests in Wales caused all the females to come off their nests and stand watching it until it left. Harriers are not yet a problem to nesting Lapwings in Britain but may become so locally if the recent increases in Marsh Harriers continue. Dyrcz & Witowski (1987) found harriers to be the main predators of waders' eggs in the Biebrza Marshes in Poland. Again active group defence reduced nest losses by about 75%. Liker (1992a,b) noted that Marsh Harriers were important nest predators in Hungary and Heim (1978) found that Black Kites were significant in Switzerland.

Seymour *et al.* (2003) found marked variations in predation rates in different sites despite the presence of a wide range of predators. They suggested that the territorial behaviour of the predators themselves limited the Lapwings' exposure to them.

After crows the most serious predators of Lapwings' nests are probably foxes and mustelids, although Badgers and Hedgehogs are probably also important. Although the evidence is conflicting, Red Foxes are normally considered the most important in Britain but this may be simply because their activities are more visible. Mink and Polecat were found important in Denmark (Iversen 1986). Mammalian predators are considerably more dangerous to adult Lapwings than crows and Lapwings' reactions to them differ sharply. This has been studied in detail for foxes by Elliot (1985b) and Seymour *et al.* (2003). The main reaction of Lapwings to foxes is to circle above the predator calling (Figure 11.7). This apparently ineffectual response serves two purposes, warning of the predator's presence and making the nests more difficult to locate, because nests without adults carry little scent and detection distances by foxes are therefore very short (Seymour *et al.* 2003). Seymour *et al.* concluded that foxes were relatively unimportant nest predators of Lapwings for several reasons. Nest predation on their site was negatively density dependent, which is typical of crow, not fox, predation. Foxes' search effort, at 57 seconds per hectare of Lapwing habitat, was too low to lead to high predation rates and high nest success was sometimes achieved in areas with foxes. They argued that foxes are much more significant predators of Lapwing chicks, which they can hunt using aural clues because chicks may continue to call despite warning calls by adults. Whether all observers would agree is doubtful. In particular, nest-camera work by the RSPB is starting to show that foxes and other mammals may be more important nest predators than crows (M. Bolton pers. comm.) and similar recent work in The Netherlands has also shown that mammalian predators, particularly foxes, seem to be the major predators of wader clutches generally there (Teunissen *et al.* 2005). M. O'Brien (in Taylor & Grant 2004) found that declines on his upland sites were more closely associated with fox abundance than habitat loss and that predation rates there correlated positively with fox abundance.

How far nest predation has contributed to modern declines in Lapwing populations, particularly in grassland, seems uncertain. Recent work in The Netherlands, for example, suggests that chick predation is a more important factor in depressing Lapwing populations (Teunissen *et al.* 2005). Chamberlain & Crick (2003) noted that predation of Lapwings' nests had increased in the UK in the 1990s and suggested that this may have contributed to recent population declines. However, Leech & Crick (2005) reported consistently higher nest success since 2000, which may be due to a decline of pairs breeding in marginal situations as the population declines. The trend towards breeding in larger groups noted by the 1998 Lapwing survey (p.115) might have contributed to this, as birds breeding in such groups are better able to protect their nests from predators (p.162).

Attempts to measure the impact of predation by predator removal experiments have not proved successful, principally because removing one suite of predators, such as crows, tends to result in other predators moving into the space thus created (e.g. Parr 1993, Bamford 2002, Fletcher *et al.* 2005). Nest predators, particularly foxes and crows, have certainly increased (e.g. Gregory and Marchant 1996, Tapper 1992) at least partly because agricultural intensification has encouraged them, through the increased availability of food sources such as sheep carrion and, for corvids, the creation also of highly favourable feeding habitats in tightly grazed grassland (Fuller & Gough 1996, Barnett *et al.* 2004). Effectively the lack of habitat diversity resulting from much modern farmland management encourages generalist predatory scavengers such as crows and foxes, which can more readily coexist with Man, at the expense of more specialised species.

The effects of modern grassland management and predation interact. As noted above the former breaches the Lapwing's natural defences against nest predation but the main thrust of that management is to promote earlier and more rapid plant growth and to raise stocking rates. Both have the effect of curtailing the nesting season and therefore the number of nesting attempts possible. Modern management of arable crops, particularly cereals, has similar effects (Chapter 4). Irrespective of whether the predation risk to individual nests has changed, under modern management regimes predation becomes more likely to prevent successful nesting because repeat clutches are less frequent (Whittingham & Evans 2004).

Modern farmland management is therefore undermining the Lapwing's third, and possibly most important, line of defence against nest predation, the ability to replace lost clutches readily (see also Figure 11.5 above).

An unwitting and unwelcome experiment on the impact of nest predation has arisen with the introduction of Hedgehogs onto South Uist, Outer Hebrides, in 1974. For whatever reason this was done, the result was to introduce a new egg predator into the most important wader breeding area in Britain. Nest predation by Hedgehogs has become a significant new cause of clutch failure there (Jackson & Green 2000). Jackson & Green also examined the nutritive value of waders' eggs to Hedgehogs and concluded that they contributed only a small proportion of energy needs. Thus it is likely that Hedgehog predation of waders' nests will be density independent, making local extinctions from this cause a distinct possibility. Study of recent population changes in the Outer Hebrides supports this, showing marked decreases in areas occupied by Hedgehogs but increases where they remain absent; Lapwings and Redshanks were the species most affected. This difference could not be explained by any other factor (Jackson *et al.* 2004). Some attempt to remove the introduced Hedgehogs is being made.

CHAPTER TWELVE

Rearing the chicks and fledging success

Lapwing chicks are delightful little creatures. The down of the upperparts and head is buffish, yellowish or sandy brown variegated and streaked with black, with a black border to the hind crown. The nape and underparts are white with a dark breast band. The bill is short and dark and the legs rather noticeably stout for such a scrap, grey or blue-grey or greyish pink in colour. The white nape is very conspicuous when chicks are active, which perhaps acts as a signal to help parents maintain contact with the brood.

When alarmed, or in response to parents' alarm calls, chicks crouch motionless with their legs drawn into their sides and head down, in which attitude the white nape patch is largely concealed. Nethersole-Thompson & Nethersole-Thompson (1986) noted that small chicks often rise and run away before danger has passed. Older chicks crouch for longer but then will also run away before crouching again.

Spencer (1953) remarked that they did this to find better cover if danger was not obviously pressing. Nevertheless, such behaviour may make them particularly vulnerable to Kestrels, which hunt them from a height and may take them over considerable distances (p.163): I have measured distances of 300–400 m over the ground from below a Kestrel's hovering stance to the kill. Seymour *et al.* (2003) noted that chicks may also continue to call despite adults' warning calls, a factor which aided foxes in hunting them and sometimes allowed the observers to locate a brood. They remarked on the foxes' highly convoluted track when searching for chicks and I have watched the same behaviour, with the fox working with its head up, suggesting that it might have been listening rather than following a ground scent.

Chicks can run soon after they are dry and they have been recorded swimming within 24 hours of hatching. If undisturbed, chicks may stay in or near the nest being brooded for some hours before being led away by the adults and the Nethersole-Thompsons record a case where the chicks were called back to the scrape in severe weather and brooded there for 41 hours. If hatching is completed late in the day, chicks may be brooded in the nest scrape overnight.

PARENTAL CARE: LEADING THE BROOD

Lapwing chicks feed themselves from hatching and the adults' role comprises leading chicks to good feeding areas, brooding them by night and at intervals by day and guarding them from predators. The female takes the major part in brooding and tending the chicks and the male in guard duties. Chicks are brooded at night for at least their first 14 days.

Broods hatched in pasture usually remain within the natal territory or its vicinity for rearing, generally moving no more than about 60–100m from the nest site over the fledging period (Spencer 1953, Redfern 1982, Milsom *et al.* 2002). However, broods hatched on tilled land, on poor feeding sites such as blanket bog or on sites where the vegetation has grown too high are moved, sometimes substantial distances, to better feeding sites, usually well grazed pasture. Damp areas or the vicinity of rills, pools and puddles are always sought if available and, even in grassland, pairs may lead chicks considerable distances to them (pers. obs.) Matter (1982) found that chicks on his arable site in Switzerland, where grassland was not available, were taken to field margins where prey were more abundant; wet soils were also more fruitful sources of food than dry. Henderson *et al.* (2004a) also found that chicks hatched in tillage in eastern England, where grass fields have become increasingly scarce, were taken to grassy field margins and to set-aside fallows for rearing (Plate 19). In coastal farmland in Sweden Johansson & Blomqvist (1996) found that chicks were led principally to seashore pastures and the open shore. They found that the mean distance moved by 35 broods from nests on tillage to their first foraging point on pasture was 99m (range 7–332m) and the mean distance to the farthest point observed was 203m (n = 25, range 61–386 m). The longest journey they recorded was of 924m. In Sussex I found that distances of 100–200m were typical when moving from natal tilled fields to rearing sites on pasture. Even day-old chicks may be moved substantial distances. The Nethersole-Thompsons quote two cases of such chicks being moved 600m and 700m, the latter swimming a canal in the process, and one case of chicks being moved 1,200m in their first three days.

It is important that adults can lead chicks to good feeding grounds quickly. Galbraith (1988a) found that proximity of pasture feeding areas to nest sites, and therefore short journeys, was a critical factor in the fledging success of pairs nesting on tilled land: 21.4% of chicks hatched in cereals with direct access to pasture fledged compared to only 8.7% of those which had to cross intervening cereal fields or hayfields. Lapwings tend to avoid nesting on tilled fields which do not abut pasture (p.54). For pairs nesting in grassland habitats such problems are much less severe. Even if parents do move their chicks significant distances, the latter can move more slowly, feeding as they go. The call and behaviour of adults moving chicks are quite characteristic, the call being the soft 'pee-wi' note used as the all clear, continuously repeated whilst the adults circle over the brood or land in front and call them on.

Chicks may not only be led considerable distances but also over considerable obstacles. In Wales I have watched them being encouraged to cross dense stands of Purple Moor Grass and soft rush and, in Sussex, some of my broods certainly passed through thorn and bramble hedges to get to pasture, although many could use conveniently placed gates. I once watched a brood of four there taken across the main drainage stream of the farm, which was about 4.5m wide enclosed by banks about 2m high, overgrown with rough grass. The female stood on the opposite bank calling continuously, whilst the male circled back and forth. Eventually the chicks swam across one after another, struggling up over the bank and out to the pool which was their destination. Crossing such obstacles is perhaps not unusual and Galbraith (1988a) recorded chicks crossing the River Forth on his arable study area. Spencer (1953) gave some interesting details about how Lapwing chicks cope with dry stone walls in the Pennines. Chicks certainly crossed these for he found broods he had ringed on one side of a well-built two metre wall were on the opposite side a few days later. In at least some cases it was very likely that the female had lifted the chicks over and there are well attested records of Lapwings carrying chicks over obstacles. Thus Ash (1948) recorded an adult stepping over a chick, holding it between its legs and flying off with it over a wide ditch, although it dropped it after a few yards. A note in *British Birds* (40: 384) also recorded a case in Norfolk in 1946 and a fairly certain case was noted in the *Zoologist* in 1892. Spencer referred to two other cases and observed that a complete brood seldom succeeded in negotiating a wall and those chicks left behind were at very high risk, although the female might go back and forth to brood. He also recorded a male taking over and independently rearing part of a brood which had become split.

PARENTAL CARE: BROODING THE CHICKS BY DAY

Very young Lapwing chicks cannot regulate their body temperature properly and are brooded at night and at intervals by day. Diurnal brooding behaviour was studied in detail by Beintema & Visser (1989b). They found that brooding was by females for over 96% of brooding time in the chicks' first 14 days but the male's share increased thereafter, which enabled the females to leave the rearing area and feed elsewhere. Brooding may be initiated by the parent walking round the chicks and adopting a brooding posture or by the chicks calling and walking to the parent, which again adopted a brooded posture. Chicks were mostly brooded together and brooding stopped when either the chicks emerged to feed or the parent rose and walked away. However, I have several times observed that parents may cover part of a brood while the others are feeding, and then call the latter to them, releasing the first chicks to feed.

How frequently and how long chicks are brooded depends to an important extent on weather conditions. Low temperatures or rain mean that chicks must be brooded more to maintain their body temperature. Beintema & Visser in fact observed that there was a threshold temperature above which no brooding occurred, which declined with age and below which bouts of foraging and brooding alternated. They showed that brooding might occur with chicks as much as 22 days old at temperatures below about 11°C. The balance between brooding and foraging changed as temperatures fell, with foraging time declining and brooding increasing. At temperatures of 5°C even half-grown chicks spent only 35% of their time feeding and young chicks (aged 1–3 days) no more than 10–20%. Rain has an extra cooling effect, increasing the need for brooding at low temperatures, an effect which also decreases with the age of chicks. Young Lapwings become completely independent of brooding at about 24 days.

Guard duties are primarily by the male, who usually stands some little way apart from the brood in a position providing a good field of view. Beintema & Visser noted that other birds and mammals, including other Lapwing families, were kept at least 20m away from the chicks and were chased by the male and by the female if necessary. Mustelids were mobbed by all the breeding waders present. Encroaching Lapwing chicks were liable to physical assault, although occasionally families were allowed to mix and, in these cases, chicks might then change families. The reaction to human intruders is different, the adults circling and alarm calling, but Spencer (1953) noted that dogs provoked the most violent hostility and mobbing, which could drive them away yelping. Adults take very little notice of grazing livestock (Spencer, pers. obs.).

The majority of broods are reared by both parents but polygynous males may tend only one brood, leaving their other females to rear their broods alone (Blomqvist & Johansson 1994). These authors noted that single parenting also arises where females lay a second clutch, leaving the male tending the first brood as they incubate it. The desertion of a brood by one adult before fledging has been widely noticed. Blomqvist & Johansson noted this in 28% of the 28 broods they studied and it was only partly due to females laying a second clutch; single parenting was almost exactly divided between males and females. The lack of success enjoyed by pairs attempting to raise a second brood (p.120) suggests that single parenting is not a particularly sound option in rearing young, probably because excluding predators becomes more difficult.

Tending the brood involves energy costs to the adults because time spent guarding and brooding chicks cannot be used for feeding. Beintema & Visser (1989b) noted that, as broods grew more independent, off-duty adults left the brood territory to feed elsewhere.

FOOD AND FEEDING BEHAVIOUR OF CHICKS

Lapwing chicks take a wide variety of soil, surface-living and aquatic invertebrates. They have to learn to feed themselves, probably mainly by trial and error. They retain more than 50% of the initial yolk contents of the egg as a reserve in the first days after hatching, whilst they do so. Large eggs have greater yolk, lipid and energy content. They result in larger chicks at hatching, which have better yolk and energy reserves for surviving the early stages of development and they grow better than chicks hatched from small eggs (Galbraith 1988b, Blomqvist *et al.* 1997). Blomqvist *et al.* also conducted cross-fostering experiments, exchanging clutches of large and small eggs. In exchanged clutches there was no difference in chick survival in clutches of large and small eggs. They concluded, therefore, that egg size did not affect chick survival independently of parental quality. Fledging success increased with the age, experience and body mass of females. Hegyi & Sasvári (1998) also observed that females laying and hatching large eggs were more likely to fledge the young hatched.

The female, having led small chicks to a suitable feeding area, tends to stand fairly still and alert whilst they forage close around her, a behaviour not unlike that of domestic chickens, and she calls them to her if they stray too far (Spencer 1953, pers. obs.). Older chicks range more widely. Chicks are able to take fast moving prey such as beetles and have been caught with butterflies in their bills (Cramp & Simmons 1982). Their foraging behaviour resembles that of adults, moving and scanning, and Boyle (1956) recorded a brood using the pattering action when less than two weeks old. This brood regularly took tadpoles, wading into a pool and catching them with a quick stab of the bill, then beating them on the ground once or twice before swallowing them: only one or two were taken at a time. Foraging rates may decline with increases in sward height, either because prey becomes less accessible or because, having rather short legs, the chicks' mobility is restricted (Devereux *et al.* 2004) but I find they also happily feed round the margins of pools and flashes in quite tall vegetation and feed along the edges between short and longer patches of growth. One advantage of such mosaics is that the taller patches offer greater security to chicks, which they certainly exploit. Food is also taken from the vegetation. In The Netherlands Beintema *et al.* (1991) classified the components of the Lapwing chicks' diet they observed by habitat. The habitat of 17 taxa out of 27 was listed as vegetation although many items were taken infrequently and a smaller range of soil invertebrates and those of dung were much the most frequent prey. Shallow pools, which are fruitful sources of insect larvae, particularly of Chironomidae, are also important chick-feeding habitats.

In pasture chicks feed much around cowpats, taking the larvae of dung beetles and flies. These are an important resource (Beintema *et al.* 1991) but the modern use of avermectins and related products to worm cattle has the potential to seriously undermine this food source, as they have marked effects on the insect fauna associated with animal droppings (McCracken 1993, McCracken & Foster 1993). I have never observed Lapwing chicks exploiting sheep droppings as they do cowpats nor have I found such records in the literature.

Figure 12.1 summarises the diet of chicks recorded in seven study areas in four European studies for which the percentage of total items taken was given, mainly in farmland but including one area of grazed saltmarsh. Coleoptera, mainly adult beetles, were much the most frequent items found but earthworms and marine worms (Nereidae), although accounting for fewer than 10% of items by number in these studies, were the most important by weight. Other important items were tipulids and other Diptera, particularly their larvae and, in Sweden, Crustacea from freshwater and saline habitats. Only Coleoptera were found in all samples. Otherwise Mollusca, Lumbricidae, Aranea and Diptera were found in half or more.

Beintema *et al.* (1991) analysed the percentage frequency of prey items in the chicks' diet in agricultural grasslands in The Netherlands. Their results are

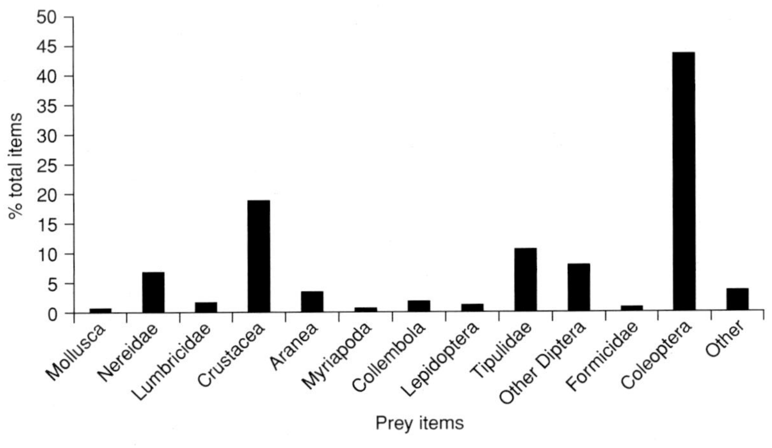

Figure 12.1. Summary of the diet of Lapwing chicks in seven European areas by percent of total items recorded. Lepidoptera were larvae and Tipulidae, other Diptera and Coleoptera include adult insects and larvae. Sources: Matter 1982, Galbraith 1989a, Baines 1990 and Johansson & Blomqvist 1996.

summarised in Appendix 3 but Lumbricidae, found in 81% of samples, and Coleoptera, found in 92%, were the most frequent items. They also found that the food of different species of wader chicks showed important differences which largely reflected the source of prey items. Lapwing chicks specialised most on species living on the surface of or just beneath the soil or on the fauna of cow dung. Lapwing chicks also showed a marked selection for prey size, the proportion of animals of <4mm taken being well below the proportion present in the habitat.

Outside these factors, regional variations in the chicks' diet tend to reflect what is available in the habitat occupied, although in arable habitats earthworms may be particularly important. Sheldon (2002) found that chicks with higher numbers of earthworm chaetae in their faeces were in better body condition than those with fewer. Seasonal changes also occur. Ausden *et al.* (2003: summarised in Appendix 3) studied feeding behaviour and diet in wet grassland in Kent and found that soil invertebrates declined in importance in the chicks' diet as the season progressed, whilst surface-active species increased; aquatic species increased very sharply in importance in June.

Beintema *et al.* (1991) found that earthworms increased in the diet as chicks grew and could obtain them more readily. They provided some evidence that insects may be insufficient to sustain chicks approaching fledging, so that a switch is necessary. This is particularly likely to be so where grassland management intensifies. High levels of fertiliser usage, especially of nitrogen, which are typical in intensively managed grassland, have a significant impact on both its vegetational structure and invertebrate populations. Several invertebrate groups of importance to chicks show moderate to severe population declines with increasing inorganic fertiliser applications (Edwards 1984, Vickery *et al.* 2001). H. Seipel (in Beintema *et al.* 1991) showed that the average body mass of insects present in grassland declined as nitrogen applications increased and that this effect emerged at levels of as little as 50kg/ha per year. It becomes very marked at levels of 200kg/ha or more, which are commonly applied to silage grass (e.g. Chalmers & Leech 1986). Earthworms may be affected at high fertiliser rates but tipulid larvae are not (Vickery *et al.* 2001) and Beintema *et al.*(1991) noted that these groups, and stratiomyid larvae which live in cowpats, become more frequent in the diet in intensively managed grass. However, the first two may become increasingly difficult for chicks to obtain as the season progresses and soils dry out and tipulids hatch, a process accelerated by drainage, an inevitable feature of grassland intensification. Beintema *et al.* concluded that intensively managed improved grassland may become unsuitable for rearing chicks as a result of these factors. Nonetheless Baines (1990), examining the impact of grassland improvement on wader populations in the English uplands, found that although improvement altered the spectrum of invertebrates available, there was no evidence that this affected fledging rates.

Nevertheless earthworms are not essential to growing Lapwing chicks in all habitats. In their study in a wet grassland area in Kent, Ausden *et al.* (2003) found that dipteran larvae, which are nutritionally rich (Byrkejedal & Thompson 1998), dominated the diet in normal years: they were taken from wet mud and the margins of shallow water, where they occurred in high densities as pools dried out. However, in 1994, a wet summer with high water levels, chicks switched to earthworms and tipulid larvae which became more readily available. There was no evidence that these dietary differences affected their growth or condition.

Drinking water is also essential to chicks and Beintema *et al.* (1991) noted evidence that food is more difficult for chicks to digest when drinking water is in short supply. They calculated that the daily water intake in Lapwing chicks near fledging was 75g per day, of which 36g derived from the water content of insect food. Earthworms and leatherjackets will supply more water but become less available in dry conditions. Dew and raindrops perhaps provide an important source. Beintema *et al.* noted that chicks raised in captivity stopped feeding entirely when drinking water was not readily available. Drought thus reduces breeding success, as it also increases adult mortality (Voous 1962).

GROWTH AND DEVELOPMENT OF CHICKS

On hatching, Lapwing chicks have relatively large eyes, and their heads, bills and, particularly, their legs are well developed relative to the rest of their bodies (Galbraith 1988d). Rapid leg and particularly head/bill growth continues during the first 12 days of the fledging period. Bill length at hatching averages 10.6mm and at fledging 22.4 (Beintema & Visser 1989a). Half that growth is achieved in the chick's first 12 days. Head and bill growth combined shows a similar pattern but leg growth is uniform throughout the fledging period (Fuller 1983, Galbraith 1988d). At hatching the wings are poorly developed but their growth rate accelerates from about 12 days old, accompanied by a rapid increase in body weight (Galbraith 1988d). Redfern (1983) noted that the remiges start to appear when weight reaches 60g, which agrees well with Spencer's timing of about 15 days (Spencer 1953). Fuller (1983) gave a similar timescale and also observed that this coincided with the growth of the bill and tarsus reaching about 60% of their development. Tail feathers start to emerge at 28 days (see Plate 15). Chicks can fly when they reach *c.*70% of the adult's weight (Jackson & Jackson 1975, Fuller 1983, Redfern 1983, Galbraith 1988d).

As Galbraith pointed out, these patterns of growth and development are adaptive and reflect the ecological demands and pressures of the fledging period. On hatching chicks have to be able to feed themselves and to follow their parents, often significant distances, to suitable feeding grounds. So structures for these functions have priority. Once settled and feeding properly wing development becomes an increasing priority as, ultimately, flight is needed to avoid predators.

Every study that has examined the growth pattern of Lapwing chicks has found that weight gain in the first 2–3 days after hatching is small or, more usually, negative. In this period chicks need frequent brooding and often have to follow their parents to feeding sites, which restricts feeding time. They draw on retained yolk reserves (see above) and their decline in weight at least partly reflects the consumption of these resources. Once they reach suitable feeding areas weight loss is quickly reversed and most graphs of growth show rather uniform daily weight gain from the end of the first week to about day 35.

Growth rates measured by weight gain vary with habitat (Figure 12.2). Baines (1990) examined growth on upland pasture over 1985–87 and, although little overall difference emerged between unimproved and improved pastures, growth was significantly poorer in improved pastures in 1985, which was a dry spring. Galbraith (1988d) found rather similar rates of growth in his Scottish areas, where chicks were largely raised on pasture, but chicks which remained on cereal fields there suffered continual weight loss although they still increased in size. This meant rapid loss of condition and high mortality. Galbraith noted that, in this context, the habitat mix was important, as the size of adjacent fields and the cropping pattern determined how readily chicks reared on tillage could reach pasture. Modern trends to field enlargement and continuous arable cropping must undermine the Lapwing's ability to fledge chicks in arable habitats.

Weather conditions have an important bearing on growth as young chicks cannot regulate their own body temperature adequately and because of their impact on the amount of time chicks can spend feeding or must be brooded to maintain body temperature. Small chicks cannot maintain their weight if foraging time falls below 25–30% of the time available and their growth is impeded below 50% (Beintema & Visser 1989b). In prolonged poor weather chicks may die of starvation. For chicks up to at least 15 days old Beintema & Visser (1989a) found a positive relationship between weight gain and the number of dry hours per day with temperatures >15°C, when the need for brooding declines or disappears. They observed that most Lapwing chicks were hatched when the number of hours

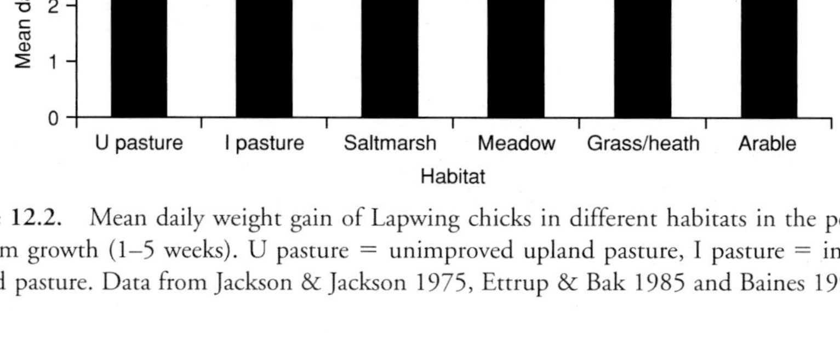

Figure 12.2. Mean daily weight gain of Lapwing chicks in different habitats in the period of uniform growth (1–5 weeks). U pasture = unimproved upland pasture, I pasture = improved upland pasture. Data from Jackson & Jackson 1975, Ettrup & Bak 1985 and Baines 1990.

meeting this criterion exceeded 25% of the hours available, which was the minimum level for survival and starting growth. Because of these processes, chicks hatched early in the season may initially grow more slowly than late-hatched chicks but gain weight more quickly later when prey such as earthworms are still accessible. Chicks hatched later show the reverse pattern, although Galbraith (1988d) found that condition at hatching did not differ. He also found that body size, including weight at hatching, affected subsequent body condition for at least 16–17 days, large chicks surviving better (above).

Chicks fledge at around 35 days but do not become independent of parents for some six or seven days after they can fly, after which they are absorbed into post-breeding flocks (Spencer 1953).

FLEDGING AND PRODUCTIVITY

Lapwings need to rear between 0.8 and one chicks per breeding pair for populations to maintain their numbers. Peach *et al.* (1994) found that steadily improving adult survival has resulted in this critical level falling from 1.13 young per pair in the 1950s to 0.83. A re-analysis of the same data by Catchpole *et al.* (1999) raised adult life expectancy further, so that this threshold is now lower.

Teunissen *et al.* (2005) estimated that 60–75% of all wader chicks lost in their study were taken by predators, 5–15% were killed by agricultural activities and 10–15% died from other causes, which included drowning or getting trapped in canals and ditches. Mowing was the most frequent cause of loss to agriculture and affected Black-tailed Godwits more than Lapwings because of habitat differences. Most losses to mowing occurred in chicks <10 days old. Although Stoats took 15% of the chicks lost to predators, in total more were lost to avian predators than to mammalian ones: Common Buzzards (12%), Grey Herons (8–18%) and crows (6%) were the species most frequently recorded.

Studies that have examined fledging success in Lapwings have generally found that most chick mortality occurs in the first ten days after hatching when, as discussed above, it is linked to poor feeding efficiency and poor ability to regulate body temperatures in the chicks. Mortality thereafter declines steadily (Figure 12.3).

The figure also shows marked differences in the scale of mortality between habitats. Galbraith (1988a) found much higher mortality in his arable area, where 80% of chicks died in their first ten days, than in his rough grazing area, where losses at this stage were 50%. Baines (1989), however, although recording a similar pattern in overall mortality, found much less difference between grassland habitats and between grass and arable land.

Fledging success also varies between years, mainly with variations in weather patterns. In the New Forest, Jackson & Jackson (1975) found that very wet weather in June resulted in poor fledging success. Drought had similar effects but that was at least partly due to the indirect effects of human disturbance in what was a

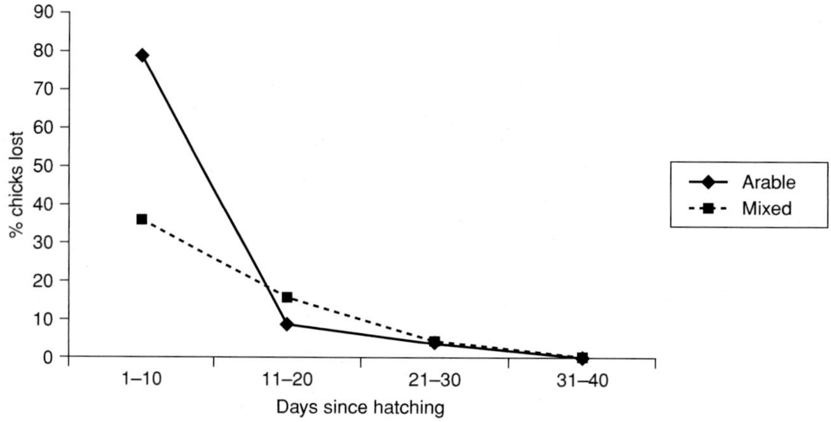

Figure 12.3. Mortality of Lapwing chicks at different ages in arable farmland in Switzerland and mixed farmland in north Germany. From Matter (1982).

holiday playground, since dry weather encouraged more visitors. Matter (1982) also found that, on tilled land in Switzerland, drought was a major cause of fledgling loss because it reduced the availability of favoured prey. Beintema (1994) also showed that the conditions of chicks in The Netherlands correlated with May rainfall; wetter Mays produced better chicks but a link with fledging success could not be firmly established from the data available. Beintema (1995) also found some indication of a seasonal variation in fledging success, with the poorest survival in chicks hatched early and late in the season.

Environmental factors such as weather have always affected the breeding success of birds but, unless they show persistent long-term trends, their impact on populations is broadly neutral; good years follow bad. Since about 1970, however, studies of farmland Lapwing populations throughout Europe have found that productivity is often below the level needed for self-maintenance. Peach *et al.* (1994) listed results from ten papers covering 24 study areas which found that only in eight was productivity at a satisfactory level, and only in four was it more than one young per pair (Table 12.1).

The table suggests clearly that trends in productivity tend broadly to decline with the scale of agricultural inputs, although recent work by the RSPB and in The Netherlands indicates that predation, particularly of chicks, is now becoming increasingly important (Chapter 11). Unimproved grassland probably remains a generally satisfactory habitat but elsewhere there is little reason to doubt that the suitability for Lapwings of improved farmland everywhere continues to decline. Many European populations must be sustained by immigration to persist.

Two distinct themes are detectable in the causes of poor productivity in these studies. Studies such as Matter (1982) and Galbraith (1988a) detected only small differences in hatching success between arable and grassland habitats but large differences in fledging success rooted in difficulties of food supply. In different

Table 12.1. The number of European Lapwing studies recording satisfactory or poor levels of fledging success by habitat, mainly since 1970.

Habitat	*Studies recording <0.83 young per pair*	*Studies recording >0.83 young per pair*
Rough grazing	1	3
Unimproved grass	1	2
Mixed farmland	4	2
Arable land	8	1
Improved grass	2	0
Fallow (uncultivated land)	1	0
Total	**17**	**8**

Data and references as in Peach *et al.*(1994), together with Imboden 1970. 0.83 young per pair is the calculated critical level of productivity required to maintain stable populations.

grassland habitats, on the other hand, Baines (1989) found that poor productivity in improved grassland was mainly driven by poor hatching success and fledgling survival differed far less from that in unimproved grass. These differences, and the present impact of chick predation, need considering in drawing up conservation measures.

CHAPTER THIRTEEN

Movements and mortality

The Lapwing is a highly mobile species and its movements outside the breeding season are complicated. Imboden (1974) noted that Lapwing populations breeding in Britain had the most complex migrations of all, with partial residents, internal movements within the country and emigration to the Continent and west to Ireland. British- and Irish-ringed Lapwings have been recovered in virtually every European country and as far east as western Siberia but over 80% of overseas recoveries of British Lapwings have been in their main wintering areas of Ireland, France and Iberia (see Figure 13.1 below). Britain, France, Iberia and Ireland are among the most important wintering areas for European Lapwings breeding further north and east (Chapter 5 and below).

Broadly speaking, Lapwings' movements are of three types, post breeding dispersal and early summer movements, migration to and from winter quarters and winter movements in response to freezing weather. It is not always clear that the first and second are separate and Hayman *et al.* (1986) noted that a westward movement begins in June and large numbers of Lapwings moult in western Europe during July to September, with onward migration continuing after the moult. Most moult on the Continent but early arrivals in Britain (see below) moult there and others cross the North Sea into Britain whilst still in moult (Appleton & Minton 1978, Appleton 2002). Appleton & Minton noted that wing moult in Lapwings is more protracted than in most waders, which may have some selective advantage, such as the need to maintain good flight capacity for summer movements, or may arise because food supplies restrict the pace of moult (G. Appleton pers. comm.). These movements, as much as cold weather movements, may partly be triggered by difficulties of food supply, with birds shifting from continental summer weather conditions (drought) to damper oceanic regimes. An extreme example of the effects of summer drought was given by Voous (1962) for The Netherlands in the very dry summer of 1959. Drought then lasted from mid-May to mid-October, resulting in a total lack of earthworms in the diet because of the hardness of the ground. As a consequence individuals collected were in very poor condition, with body weights up to 48% lower than normal and increasing numbers of dead birds were found. The overall effect was very similar to that of severe winter weather.

SUMMER MOVEMENTS

Summer movements start early. Adult and juvenile Lapwings disperse from their breeding grounds after breeding and join moulting flocks. Such birds may arrive in Britain from the Continent as early as late May, these presumably being mainly failed or non-breeders, and more extensive movements occur in June, July and August of birds breeding in the Low Countries (Belgium and The Netherlands) east to Poland, Hungary and western Russia. Ring recoveries analysed by Imboden (1974) showed that eastern European Lapwings basically move west to northwest at this time, with most recovered in the Low Countries, France and Britain. Birds from the Low Countries and Hungary were also found in Spain and Italy respectively. Scandinavian birds moved more southwest, with recoveries mainly in Denmark and some in Britain and France. Imboden's analysis showed significant differences in the proportions of adults and juveniles making long distance movements at this time, with 70% of second year birds and 77% of adults recovered within 100km of their place of origin compared to 95% of juveniles, although the difference may reflect greater juvenile mortality closer to ringing sites.

The actual number of recoveries at different sites at this and any other season must be regarded with caution, as a prime source of ringing recoveries in Lapwings and other waders is hunting (Wernham *et al.* 2002). This can bias the pattern of recovery towards those areas where hunting is most intense. The general pattern of recoveries in France provides a clear example, showing a marked concentration in the southwest and on the coast, particularly south of the River Gironde (Imboden 1974, Wernham *et al.* 2002). However, the French Winter Atlas (Yeatman-Berthelot 1991) based on records from 1977–1981, shows no exceptional concentration of Lapwings there and instead large concentrations in central and northern France, where there are many fewer recoveries. Whether or not the pattern of movements at this season illustrated by Imboden (1974) has changed is not clear, although it seems unlikely, but the scale of summer immigration into Britain, for example, has certainly declined since the 1970s (e.g. Gillings 1999), probably as a result of the general decline in European breeding populations (Chapter 3).

Further east, in Russia and Siberia, similar summer movements seem not to occur, perhaps because breeding finishes later. Dementiev & Gladkov (1969) noted that, as fledglings acquired flight ability, Lapwings formed fairly large flocks which roamed about the fields and meadows, indicating a nomadic habit rather than any definite pattern of movement. Wetland habitats may be important for they also noted that migrant flocks feeding on the steppe flew to water two or three times daily.

AUTUMN PASSAGE

Autumn passage proper begins in September and movements continue through much of the autumn and winter. Movements of the British population have recently been analysed by Appleton (in Wernham *et al.* 2002). He divided the population into four subgroups, from southwest and southeast Britain (south of 54°0′N and east/west of 1°30′W) and northwest and northeast Britain (north of 54°0′N and basically the eastern lowland zone and the western upland zone). The pattern of recovery of birds from these sub-populations in winter is summarised in Figure 13.1.

About 20% of wintering British Lapwings have been recovered within Britain. Otherwise Lapwings from northern Britain move particularly to Ireland and to a

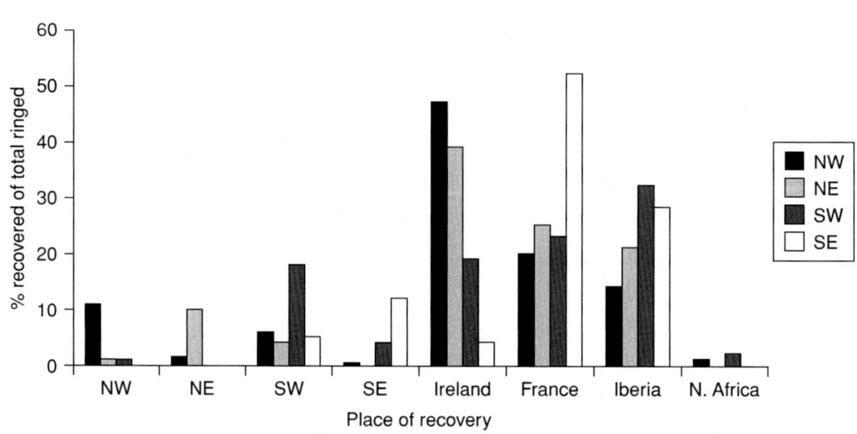

Figure 13.1. Approximate percentage of ring recoveries from different British Lapwing subpopulations recovered in different wintering areas; <1% of birds ringed in the southwest were also recovered in The Netherlands. Source: Wernham *et al.* (2002).

lesser extent to France and Iberia. Birds from southwest Britain move primarily to France and Iberia, with significant movements also to Ireland. From southeast Britain the main movement is to France and, to a lesser extent, Iberia. In both France and Iberia the recoveries are strongly biased towards the coast but see above for reasons to regard this pattern with caution.

Imboden (1974) noted a similar pattern of movements by British-breeding Lapwings and also separated birds by age. This analysis showed that significantly more juveniles emigrated to the Continent, being found >700km from source, and more adults remained in Britain and Ireland and were recovered between 100km and 700km from where they were ringed.

Movements of winter visitors into Britain start in September and the WeBS counts now show an annual peak between November and January, occasionally in February. Figure 13.2 plots the mean monthly totals for Britain for 1970–75 and compares these with the monthly means for 1995/96–2001/02. Only the coastal counts are shown as only these were made in the earlier period. Apart from the considerable increase in numbers shown (discussed in Chapter 5) the counts in Figure 13.2 suggest a change in the pattern of arrivals, with a peak in January consistently noted until the mid-1970s but peak numbers now usually occurring in November/December and numbers remaining high until February. Departure is now very rapid in March.

Although the inland counts cover only selected wetland sites and are not complete population counts, the WeBS counts for all coastal and inland areas combined (Figure 13.3) suggest that, despite annual fluctuations, the population present from November to February shows a strong underlying tendency to stability. This perhaps arises because there is less onward movement in the late autumn and early winter in the milder winters prevailing since the early 1990s. Certainly the number of recoveries of British Lapwings shot in France and Iberia has declined

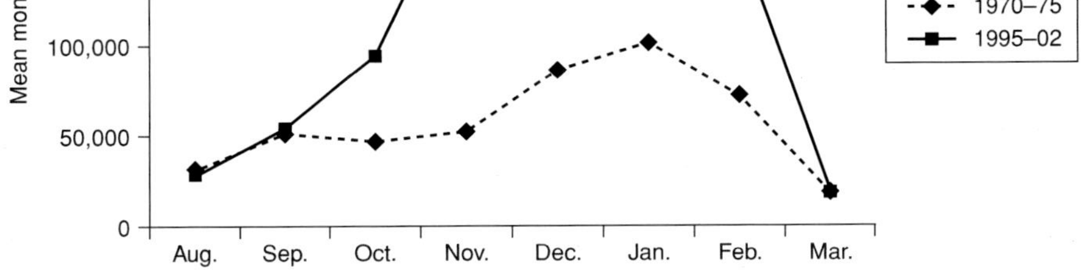

Figure 13.2. Mean monthly totals for coastal counts of Lapwings from August to March for 1970–75 and 1995–2002. Source: Estuary Counts and WeBS Reports.

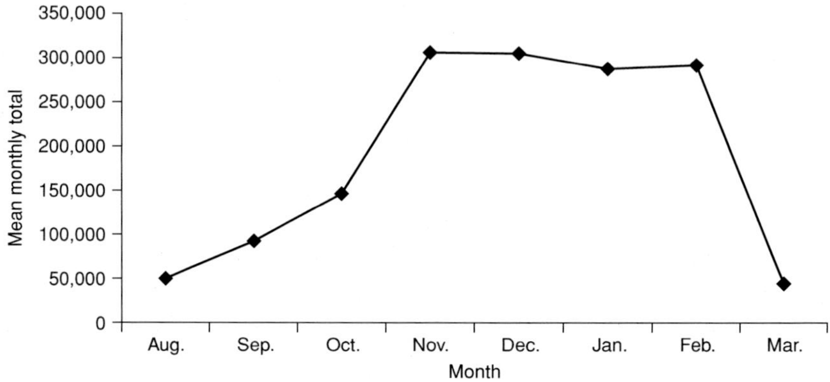

Figure 13.3. Mean monthly counts of Lapwings from August to March in all areas covered for WeBS counts in Britain during 1995–2002. Source: WeBS counts.

(Peach *et al.* 1994) which is most likely to reflect a decline in the number of birds going there and therefore vulnerable to hunters. Comparison of Imboden's (1974) results for the Low Countries with the 1987 Dutch Atlas also suggests that something similar has happened in The Netherlands. For Imboden noted that no more than 5% of recoveries of Dutch ringed birds up to 1969 were found in the breeding area in January and only four recoveries there were for populations further north and east. Although van der Winden *et al.* (1997) noted that almost all Lapwings left The Netherlands during prolonged cold spells (see below), the Dutch Atlas (SOVON 1987) nevertheless noted that 61% of 5km-squares then held Lapwings in January.

Nevertheless there are also differences in the timing of peak numbers in England, Wales and Scotland. In Scotland peak counts mainly occur during October–December, in England they fall mainly in November–January and in Wales especially in December (Figure 13.4). These differences suggest that significant internal movements occur during the winter.

Analysis of European ring recoveries by Imboden (1974) showed that migrant populations wintering in Britain and Ireland derived mainly from the Low Countries, Denmark, Norway and Sweden (88% of recoveries), with some birds also arriving from central and eastern Europe: Ireland attracts birds mainly from Scandinavia. In contrast to British populations Imboden found no significant differences in the distances travelled by adults and juveniles on autumn passage from these populations. Appleton (2002) showed a similar pattern for Britain but with perhaps more emphasis on central and eastern Europe, which produced 14% of recoveries up to 1997, although not all these birds were necessarily present in winter. Appleton also noted that there were no recoveries from Ireland into southern Europe, which may derive from a lack of winter emigration, such as occurs in Britain, or from a lack of ringing. Hutchinson (1989) noted that

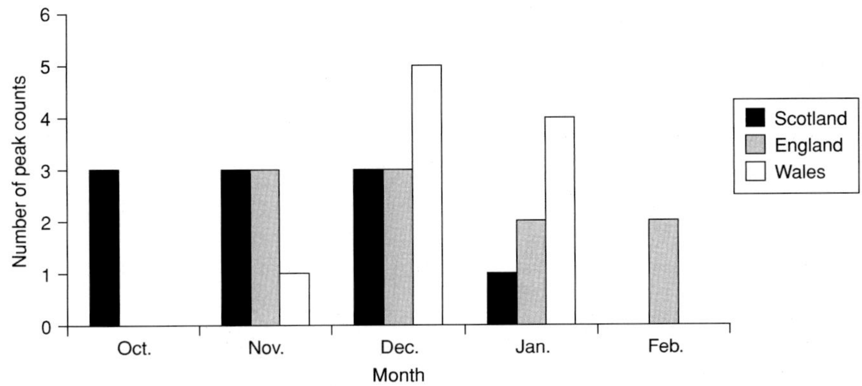

Figure 13.4. Number of peak winter counts of Lapwings in each winter month noted by WeBS from 1992/93 to 2001/02 in England, Wales and Scotland.

nothing was known of the movements of Irish-breeding birds but Dobinson & Richards (1964) noted that heavy mortality occurred in Ireland in the winter of 1962/63, as did Jourdain & Witherby (1918) for the harsh winter of 1916/17. Such records suggest that, even in severely stressful conditions, Ireland is the end of the line.

Elsewhere, Lapwings breeding in central, northern and eastern Europe winter largely in France, Iberia, Italy and down into North Africa, mainly Morocco (Figure 13.5). Italy draws birds mainly from central and eastern Europe, particularly from Hungary, which also cross the Mediterranean to winter in North Africa east of Morocco. Some birds from these populations also winter on the Mediterranean islands.

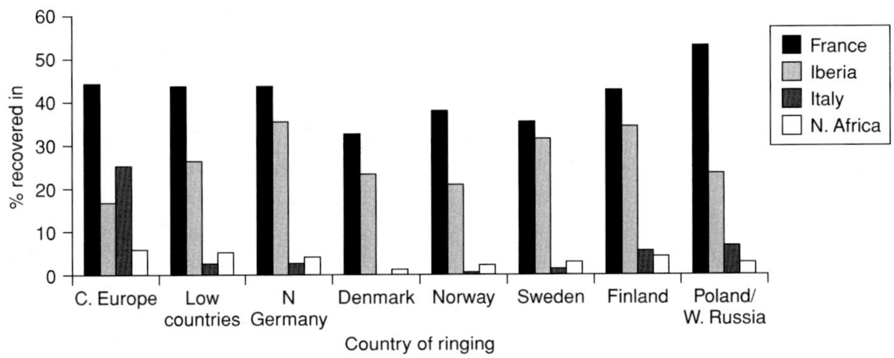

Figure 13.5. Main wintering areas of European Lapwing populations as shown by ringing recoveries. Source: Imboden (1974).

There are large variations in the number of Lapwings which reach Morocco for the winter, deriving from the varying severity of European winters (page 67). Many of these birds probably pass through Spain and Finlayson (1992) noted heavy passage from Spain into Morocco in some winters. Nevertheless there seem to be some interesting differences in the sources of ringed birds recovered in Spain and Morocco, with more from The Netherlands and Belgium in Morocco and more from Britain and Scandinavia found in Spain (Figure 13.6).

Imboden (1974) found that autumn passage from central Europe occurs mainly in November and from Scandinavia in October or earlier. The Netherlands are an important staging post for Lapwings moving to wintering areas further west and south. Van der Winden *et al.* (1997) gave preliminary results of a country-wide survey of staging Lapwings in October and November 1996, which found totals of 954,000 birds in October and 686,000 in November. As Lapwings spend little time feeding in intertidal areas there, these counts were considered to be grand totals for The Netherlands and a peak autumn count of around a million birds was thought likely to be typical. The sharp decline noted in November was associated with a cold snap and during prolonged cold spells these authors noted that almost all Lapwings leave the country. Del Hoyo *et al.* (1996) also indicated a November peak of 100,000–200,000 birds in Denmark, most of which move on through. For example Meltofte (1980) found as few as 27 in the Danish part of the Wadden Sea in December 1978. Throughout, the basic trend of movement is strongly southwesterly and thus differs, for eastern European populations, from the more west to northwest trend of summer movements.

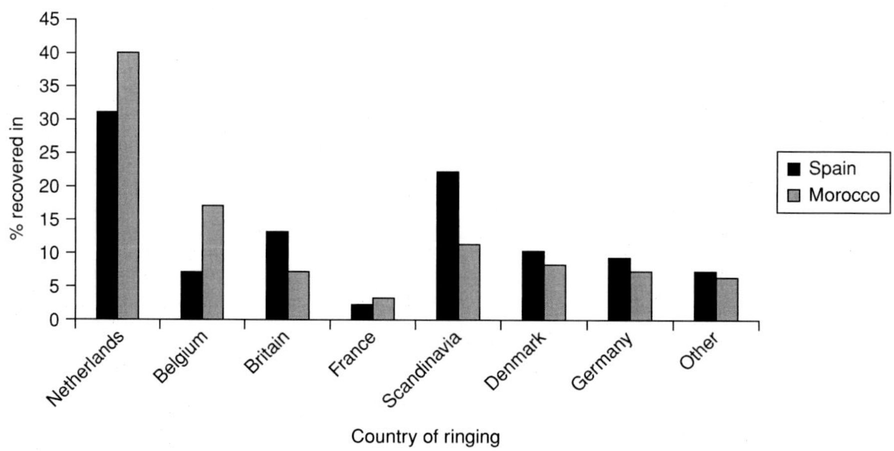

Figure 13.6. The percentage of ringed birds from different European countries recovered in winter in Spain and Morocco. The differences between the proportions recovered in Spain and Morocco from The Netherlands and Belgium and from Britain and Scandinavia are statistically significant ($\chi^2_7 = 36.65$, $p<0.01$). Source: Finlayson 1992 and Thévenot *et al.* 2003.

Further east in Russia and Siberia Dementiev & Gladkov (1969) showed that autumn departures by Lapwings occurred in the period July to September over a very wide area between 52–58°N and 25–95°E and into the Far East in Ussuriland, where they left in September. South of 50°N movements were most marked in September and October over the same range of longitude. How far the timing of this movement has changed with the expansion in range which has occurred is unclear but comparison with records given by Rogacheva (1992) for central Siberia suggests rather little change.

Ringing recoveries indicate that at least some Lapwings from Belarus, the Ukraine and Russia east to Moscow move west into European wintering areas. Further east, although birds from Omskaya (*c.*72°E) have been recovered in Britain and Italy, observations of visible passage have shown large southerly movements across the steppes and Mongolia (Dementiev & Gladkov 1969). These birds are presumably headed for winter quarters in the Middle East, India and China.

In Europe movements continue through much of the late autumn and winter period, with the ring recoveries indicating a peak arrival in Iberia and North Africa in January (Figure 13.7).

The pattern for Iberia has been recently confirmed by Asensio (1992) and Fornasari *et al.* (1992) noted that peak winter counts in Lombardy, north Italy, occurred in December. As already noted the scale of movements into southwest Europe and North Africa varies with the severity of the northern winter, birds remaining further north and east in milder winters. Thus more British Lapwings are recovered in Iberia in severe winters, for example 56 in the winter of 1962/63 compared to the normal level of five or six annually.

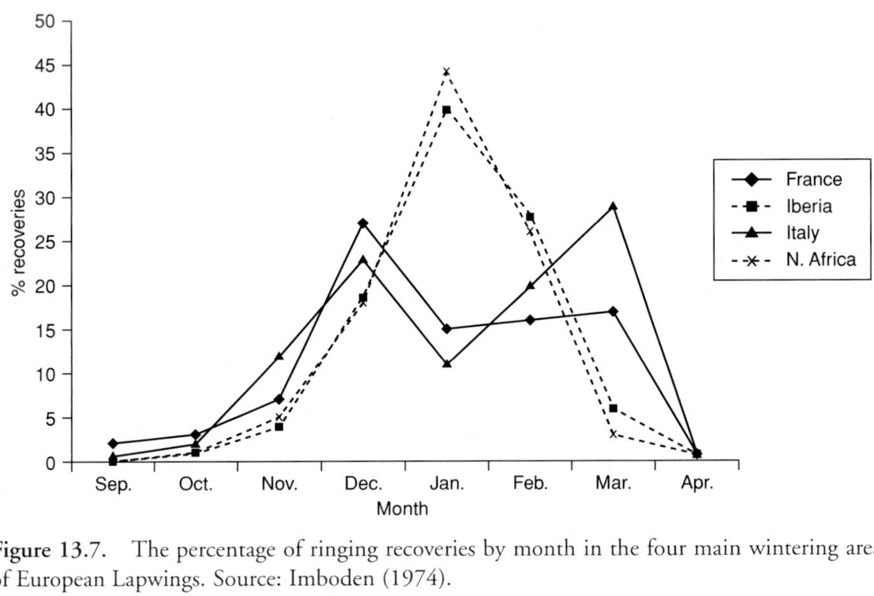

Figure 13.7. The percentage of ringing recoveries by month in the four main wintering areas of European Lapwings. Source: Imboden (1974).

COLD WEATHER MOVEMENTS

Extensive movements in response to severe cold are a feature of Lapwing movements in late autumn and winter and spectacular diurnal cold weather movements may be recorded. In the run of mild winters which began in the early 1990s these have become infrequent in Britain but they were a common feature of Lapwing accounts in county avifaunas and reports there into the 1980s. For example I found cold weather movements of 2,000–3,000 per hour on the Sussex coast not uncommon in cold snaps until the early 1970s. Movements might continue all day and involve birds arriving from the sea and moving west as well as departing out to sea south and southwest. They were often accompanied by smaller numbers of other species, particularly Golden Plovers, Skylarks, thrushes and Starlings. Such movements most commonly occur in late December and January but may do so as late as early March (see below) and the largest such movement yet noted along the Sussex coast was of 40,000 Lapwings departing out to sea at Shoreham on 31 December 1978.

The birds move on a broad front and these cold weather movements were noted right along the English south coast, at many sites inland and on the east coast. For example 20,000 Lapwings were counted flying south over Langstone and Portsmouth Harbours in Hampshire on 30 January 1972, when 5,000 also flew south inland over Fleet Pond in the north of the county (Clarke & Eyre 1993). Similar large movements were noted over Southampton and the west end of the Solent on 31 December 1978, coinciding with the largest Sussex movement noted above. Further west the trend of such movements tends to be more westerly, with Lapwings moving to Ireland and movements of up to 9,000 in a day have been observed in Glamorgan (Hurford & Lansdown 1995). Inland, in Buckinghamshire, Lack & Ferguson (1993) noted that the large wintering population of Lapwings moved off south or southwest with prolonged periods of freezing weather or snow but often returned to the same fields a week or so after a thaw. In really severe winters virtually all Lapwings may leave Britain, as they did in 1962/63 and again in early 1982 (Dobinson & Richards 1964, Kirby & Lack 1993). Such cold weather movements have become infrequent since the early 1990s.

SPRING MIGRATION

Lapwings start to return to breeding areas from late January/February in southern Europe (Figure 13.7) and spring movements take much less time than those in autumn. In Britain spring arrival on the breeding grounds generally starts in late February or early March and there seems to be little variation in this pattern over time or with latitude for, where they discuss it, virtually all the 19th and 20th century county avifaunas indicate the same general picture. Some accounts indicate that arrival in the uplands may be up to a month later but this is not invariable: Spencer (1953) gives dates between 18 February and 4 March in seven years at 1,100 ft (340m) in the western Pennines near Burnley in Lancashire. The exact timing of arrival also depends partly on the persistence of winter but even in late springs arrivals on the breeding grounds may vary by only a few days and Lapwings appear very rapidly once the frost eases. Thus in both 1956 and 1963, when hard frost there persisted right to the end of February, passage and return to breeding sites was noticeable in Sussex in the first week of March and, in 1963, the first returning birds were actually noted on 15 February, two weeks before the final thaw set in; the first in northern France was noted on the 10th (Dobinson & Richards 1964). After the exceptionally severe and snowy winter of early 1947 Spencer (1953) noted that his breeding birds arrived between 15 and 21 March, a delay of around a fortnight.

Imboden (1974) showed that the peak spring passage occurred in mid March in Switzerland, The Netherlands and in south and northwest Germany, in late March in southern Russia and northeast Germany and, by mid April, had reached central Sweden. Ringing recoveries also show a marked spring peak in Italy in March (Figure 13.7). Lapwings arrived in early March in southwest Norway (Grønstøl 1996) and, in southern Sweden, Högstedt (1974) showed a mean date of 24 March for the arrival of 23 females; males were up to 10 days earlier. In Finland Vepsäläinen (1968) noted that Lapwings usually arrived at the end of March in the south, reaching Rovaniemi, on the Arctic Circle, on a mean date of 30 April, with a range of 5 April to 14 May over 14 springs.

Dementiev & Gladkov (1969) gave a detailed timetable over the former Soviet Union, which showed a progressive pattern of arrival northeast across the breeding range. At any given latitude arrivals were progressively later with longitude. Thus at sites on latitude 55°N, for example, birds arrived in February/March at longitude 21°E, around 30 March at 32°E, in the first half of April at 40°E and in late April at 95°E, but there was markedly less delay with latitude at the same longitude. Records given by Rogacheva (1992) suggest an interesting pattern of movements in central Siberia. She noted that birds first arrived in the Tomsk area (*c.*57°N 85°E) during 4–9 April and often appeared further northeast at Mirnoye in the Yenisey valley, which is north of the breeding range at *c.*62°N 89°E, by the middle of the month but birds did not reach Krasnoyarsk, about 700km further south up the Yenisey, until later in the month. She suggested that the birds at Mirnoye overshot from the southwest rather than arriving from the south as did the birds at Krasnoyarsk.

Early-arriving groups of Lapwings in central and northern Europe regularly reverse their direction of migration as changes in weather conditions affect their advance (Vepsäläinen 1968). This author quoted the example of a reversed migration caused by heavy snow in north Germany on 30 March 1952, which involved more than 10,000 birds over a 10km front. He noted that the pioneering patterns of arrival and movement by Lapwings in Finland were very different from the movements of other early breeding waders there, such as Ringed Plovers,

Dunlins, Curlews or Green Sandpipers, whose arrival tracks the spring isotherm of 0°C and the thawing of the ground.

Occasional catastrophes occur when these early-arriving Lapwing are caught by sudden cold snaps or late snow on the breeding grounds and are unable to move away. One such catastrophe, described in detail by Vepsäläinen (1968), occurred in April 1966, when a severe snow storm over 12 to 13 April affected southern Finland, the Baltic States, southern Sweden, Denmark and northern Germany. Very low temperatures prevailed until 20 April and the snow did not start to melt until the 25. Mortality in Lapwings was very high, with at least 1,300 birds picked up dead in southern Finland and Estonia and total losses were estimated to run into many thousands. The impact on breeding populations was severe in areas where mortality was high, with the breeding population of southwest Finland and southeast Lake-Finland declining by 30–60% and in places by 90%, declines of 80–90% in Scania, southern Sweden, and of 50–90% in many Estonian breeding areas. Marked declines were also noted in Denmark. Breeding populations outside the area directly influenced by the cold were unaffected and birds were displaying on snow covered shore-meadows at Turku, southwest Finland, by 22 April.

ABMIGRATION

Abmigration is defined as an anomalous migratory movement of a particular type (Campbell & Lack 1985) and here involves spring movements by birds that had remained in their native areas during the winter. It is a distinctive feature of Lapwing migration and arises partly because there is a marked overlap and mixing of eastern and western breeding populations in winter quarters. Individuals from one population may then get caught in and migrate with flocks from a quite different breeding area. In this way Lapwings ringed as pulli in Britain have been recovered in subsequent breeding seasons in Scandinavia, Russia and west Siberia to 68°E (Figure 13.8). Dutch birds have been similarly recovered in eastern Europe and across Russia as have a few from Scandinavia. Most of these birds have been three years old or more when recovered and may have been breeding in these eastern sites for more than one year (Mead *et al.* 1968, Appleton 2002). Imboden (1974) also recorded 31 recoveries in the breeding season of west European Lapwings more than 1,000km from their natal site, the furthest of which was recovered more than 5,000km from its place of origin at 91°E. These movements presumably promote gene flow between Lapwing populations and underlie the lack of subspecific differentiation over the species' vast breeding range (Mead *et al.* 1968).

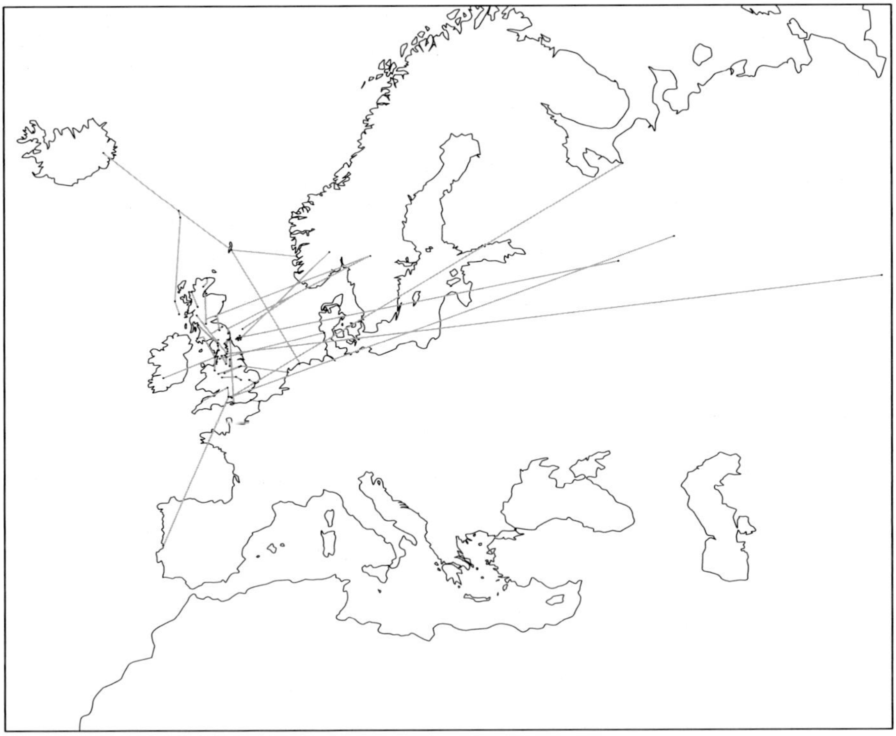

Figure 13.8. Movements of British and Irish Lapwings ringed as pulli and recovered in a subsequent breeding season when of breeding age. Only movements of more than 20km are shown (lines), with points showing recovery locations. Reproduced from Appleton (2002) with permission from the BTO.

VAGRANCY

As might be expected with so mobile a species as Lapwing, extralimital vagrancy is quite widely recorded. Thus, in the Arctic, Lapwings have been recorded as overshooting migrants in Jan Mayen, Bear Island and Spitzbergen (Cramp & Simmons 1982) and are described as a casual visitor to Greenland by Salomonsen (1967), who gave 5–6 records totalling 34 birds, all for the period October to April in the south. The source of such records is likely to be similar to that of North American records (below).

In the Nearctic it has been recorded on Baffin Island and along the east coast of North America from Labrador south to South Carolina, with records also from Ohio, New York State, the Bahamas and Barbados (Godfrey 1966, Bagg 1967, Sibley 2000). Sibley also shows records for Florida but at least some of these were questioned by Bagg, whose correspondents considered them to refer to a South

American species, the Southern Lapwing. All the records fall into the period September to March except for one on Barbados on 25 July 1963. The great bulk of occurrences are for December to February and most have been recorded in eastern Canada, from Labrador to New Brunswick. The records are dominated by two exceptional invasions, in December and January 1927/28 and during January and February 1966. In 1927 the total numbers involved were possibly as high as the low thousands and at least one flock of 500 birds was reported. Only one bird was recorded then in the United States and the highest numbers were found in Labrador and Newfoundland, particularly between 17 December 1927 and 15 January 1928 (Godfrey 1966, Bagg 1967). The birds involved were presumed to be aiming for Ireland in a period of severe weather with very strong easterly winds, missed their landfall and continued on across the Atlantic. One bird recovered had been ringed in England. The 1966 invasion was much smaller and involved *c.*33 birds in January and February, mainly in Newfoundland, with records also from Quebec, Nova Scotia, New Brunswick and the islands of St. Pierre and Miquelon (Godfrey 1966, Bagg 1967). Outside the late autumn/winter period Godfrey only quoted one record for Canada, for Ketch Harbour, Nova Scotia, on 17 March 1897.

Lapwings are much rarer in the United States, where there seems to be a total of about 14 records, usually of single birds in the northeast States. Again most records are for November and December.

To the south of the normal range Lapwings have been irregularly recorded on the west coast of Africa from Mauretania south to the Senegal Delta, although these birds may be irregular winter visitors rather than vagrants. There are two records for the Cape Verde Islands, on 23 and 24 December 1987 on Sal and on 20 February 1999 on Boavista (Clarke 2006). Clarke also noted that birds have been recorded on the Selvagens, between Madeira and the Canaries. In the Sudan it has been recorded south to Kosti, on the Nile about 250km south of Khartoum: a record for South Africa is now rejected (Urban *et al.* 1996). Further east they are vagrants to Baluchistan (Summers *et al.* 1987), vagrants or very rare passage migrants to Bhutan, stragglers to Rajastan and northern Gujerat in India and vagrants to Bangladesh (Ali & Ripley 1969, Grimmett *et al.* 1998). Vaurie (1965) noted Lapwings as vagrants to Sakhalin and the Kurile Islands.

MORTALITY

Using data from ring recoveries of birds ringed as chicks and subsequently found dead, it is possible to construct patterns of survival and mortality in adult and fully grown Lapwings. Using such ringing recoveries Haldane (1955) estimated an annual adult survival rate of 65.7% ± 2.2% on recoveries from 1909 to 1952. Peach *et al.* (1994), in an analysis of recoveries from 1930–1988, estimated adult survival at the higher rate of 70.5% ± 3.1% and noted that, from 1960, survival rates had increased to 75.2% ± 4.6%. Elsewhere in Europe similar calculations have produced estimates of adult survival of 66.8% ± 2.2% in Denmark during 1920–1978 (Bak & Ettrup 1982) and 66.9% in Scandinavia and 70.6% in central Europe (Glutz von Blotzheim *et al.* 1975). These figures suggest little variation across European populations. As most adult mortality occurs in winter and the wintering areas of European breeding populations coincide to a marked extent, this is not perhaps surprising.

As with most birds, survival of first-year Lapwings is lower, estimated by Peach *et al.* (1994) at 59.5% ± 4.% over the period 1930–1987. Mortality of first-year birds is now highest in June, July and August and there is a secondary peak in December and January.

Several important points arose out of the analysis by Peach *et al.* (1994). The increase in adult survival rates has occurred over a period when breeding populations have been declining. An adult Lapwing's expectation of life has risen from 2.4 years in 1960 to 3.5 in 1990, representing, as noted in Chapter 3, an increase of 40% in the breeding potential of the individual. The re-analysis of the same data by Catchpole *et al.* (1999) raised that life expectancy by another 13%. Peach *et al.* also noted that the decline in adult mortality lowered the critical level of productivity needed to maintain populations (p.175). In addition the survival rate of young birds reported by Peach *et al.* was similar to those found by Glutz von Blotzheim *et al.* (1975) and Bak & Ettrup (1982) and these survival rates have shown no long-term trend since 1930. Clearly, therefore, the population declines seen since the 1970s are not due to reduced survival of either adults or young birds.

Causes of death in Lapwings suggested by the ringing recoveries of British and Irish birds were hunting (57%), other human related incidents (22%), predation (12%) and disease (5%). Peach *et al.* (1994) noted that most of the 'other human-related incidents' between 1930 and 1988 were road casualties and that 7% of recoveries then were of birds which had hit overhead wires. They also observed no differences in the frequency of these causes between adults and juveniles. The ringing returns also showed that whilst 93% of recoveries in France, Ireland and Iberia were from hunting, 93% of other human related incidents came from Britain, perhaps an indication of traffic density there, and Britain also accounted for 95% of the recoveries through predation.

Although they can indicate something of the accidents that befall them, such analyses of the finding circumstances of ringed birds give little real idea of the causes of mortality because they are heavily biased towards human activity or presence. Nevertheless, over the period since 1960, Peach *et al.* (1994) noted that the numbers of Lapwings recovered as shot declined by 50% or more. Their analysis in fact showed that annual variations in the survival of both adult and first-year Lapwings were largely explained by variations in winter soil and air temperatures; mortality in both age classes increased as soil and air temperatures declined. Lapwings are notoriously susceptible to freezing winter conditions and a major cause of mortality in this species must be starvation and trauma in difficult winter conditions. Survival rates were particularly low during notably severe winters, falling to 60% of the long-term mean during 1930–1988 in 1939/40, 56% in 1946/47 and 48% in 1962/63 for example. Summer drought can also cause severe mortality, as Voous (1962) showed for the dry summer of 1959 in The Netherlands (p.179).

It is difficult to assess the significance of predation from ringing recoveries. Most Lapwings taken by predators are unlikely to be recovered and it is possible that an important source of such recoveries comprises remains found at the breeding sites of raptors, introducing a seasonal bias. It is hardly likely, however, that the modern recovery of raptor populations has much affected adult Lapwing populations in Britain at least. Adult survival could hardly have increased as it has if increased predation was exerting significant pressure on numbers. That is not to say that raptors do not kill Lapwings. Ratcliffe (1980) observed that Lapwings comprised *c.*2.5% of Peregrine kills in Britain and Newton (1986) noted that they comprised 1–2% of Sparrowhawk kills. I have also watched a Merlin take an adult out of a winter flock.

Hunting pressure and the impact of severe weather are linked because cold weather movements increase numbers wintering in southern Europe where hunting pressure is most severe. There seems little reason to doubt that the present trend to much milder winters (e.g. Figure 5.4) and the decline in hunting pressure underlie the improvement in adult survival. That this favourable regime can no longer support increases in the breeding population, as it did in the early 20th century, should be a matter of considerable conservation concern. It reinforces the importance of reduced habitat availability and declining productivity with the intensification of farming as the main drivers of population decline.

CHAPTER FOURTEEN

Conservation and the future

Despite the adaptability that Lapwings have shown in the past, there can be little doubt that adaptability to agricultural change has reached its limits with the modern intensification of farming. A species that has declined to the degree that the Lapwing has in Britain can only be considered as in serious trouble, at least as a common component of the farmland avifauna there.

This chapter on the conservation of Lapwings is largely based on practice and experience in Britain, where I have been involved to some extent in Lapwing recovery programmes. However, it should be noted that many of the problems Lapwings face in Britain, particularly the loss of wet grasslands and other habitats, agricultural intensification, the switch to increasing areas of autumn tillage at the expense of spring-sown crops, and the spread of crops such as forage maize, have arisen throughout Europe (see Chapter 3 for a detailed discussion). The difficulties posed by rising stocking rates, particularly of sheep, seem to be less of a problem in parts of continental Europe, although they have been particularly cited in Spain (Martí & Del Moral 2003). Indeed, declining grazing management has emerged as a problem in parts of eastern and northern Europe (Chapter 3). Everywhere the recovery of Lapwing populations will now rest on the types of measures described here for Britain, to restore habitat and to promote less-intensive agriculture. Consideration should also be given to providing total protection for the species in Europe (see Chapter 3) and to protecting breeding habitats outside farmland, which may be particularly valuable nurseries.

It is, I think, important to note that the agricultural changes of the 19th century in Britain, to which the Lapwing adapted, were particularly extensive habitat changes, most obviously from the Waste to rotational farming habitats. Lapwings certainly declined in this period but that was at least partly due to excessive exploitation by egging and hunting (Chapter 3) and by and large the species remained a common farmland bird in a farming system which remained broadly sympathetic to its needs (see Holloway 1996). In addition, much poorly drained grassland persisted throughout farmland during the 19th century and agriculture also went into a significant recession in the last quarter of the century, which persisted into the middle of the 20th century. These were both factors, together with the diversity of the farmland which emerged, which helped to limit the impact of 19th century agricultural change on farmland birds generally (Shrubb 2003).

Although habitat changes have played some part, particularly in wet grassland, the modern agricultural revolution has differed strikingly from its 19th century predecessor in the extent to which it has been dominated by change in farming methods and technology. These have been the main determinants of decline in farmland bird populations. It is, perhaps, the comparative subtlety of many of the changes wrought in the Lapwing's habitat by such management changes that makes it difficult for many farmers to accept that farming change is responsible for the decline of farmland birds such as Lapwings. The most visible feature of the modern agricultural revolution, which is a European phenomenon, is the steadily increasing uniformity in habitat and management that it is imposing on the farmland ecosystem.

As preceding chapters have shown, in improved grassland habitats higher fertiliser rates promoting more rapid plant growth, which inhibits nesting, increased stocking rates and lowered water tables have combined both to curtail the nesting season and to limit nesting success. Besides their exposure to increased risk of trampling by farm stock or destruction by farm operations, nests are more vulnerable to predation, clutch replacement is much reduced and grassland may no longer always provide suitable conditions for rearing chicks. In arable land Lapwings remain strongly attracted to spring tillage for nesting because, initially at least, it offers bare or sparsely vegetated land. Nevertheless the major switch to autumn cereals has limited the availability of such habitat, a change exacerbated by modern cereal management involving pre-emergent herbicides and early nitrogen applications, which promote rapid early growth (p.55). Such management changes have spread into spring cereals, again curtailing the nesting season in what was, until 1985 at least, the most successful farmland nesting habitat (p.55). Crops such as maize, peas and sugar beet are now preferred (p.54) but they are much less widely distributed. Field amalgamation and the decline of grassland/tillage mixtures have also been important. They result in the loss of chick-rearing habitat in arable land and involve chicks in longer and thus more hazardous journeys to reach suitable fledging sites. All these factors contribute to declining diversity in farmland.

Pesticide use in crops may have had a limited impact on Lapwings, although organochlorines appear to have had some effect on nesting performance (Chapters 10 and 11). Danish experience makes for an interesting comment here. Modern agricultural change there has followed a very similar course to that in Britain, with the major difference that use of pesticides and inorganic fertilisers in Denmark has been sharply reduced in favour of organic farming since 1983. Coincident with these changes populations of many farmland birds, which are declining in Britain, have remained stable or have increased. The Lapwing is an exception and has continued to decline in Denmark (Fox 2004). Fox noted that causality had not been clearly established but his results suggest strongly that, for Lapwings, the changes in the structure of farmland habitats brought about by modern agricultural intensification are particularly important. Thorup (2005) particularly stressed the decline of spring tillage there (p.30).

Biocides in the form of veterinary medicines are of greater potential significance. The organochlorine dieldrin was widely use as a sheep dip against scab until 1962, which presumably contributed to the decline in nesting performance noted in Chapter 10. It was replaced in 1962 by organophosphate compounds and more recently these have been superceded by synthetic pyrethroids, particularly cypermethrin. Both have some systemic action in the animal and have some effect on the fauna of animal dung (McCracken & Bignal 1991). Cypermethrin is highly toxic to fish and aquatic invertebrates and its authorisation for marketing as a sheep dip has recently been suspended (E. Andrews pers. comm.).

Other veterinary medicines which have a greater potential to affect species such as Lapwings are antiparasitic drugs of the avermectin group introduced in 1981 for worming cattle and sheep. These drugs are absorbed systemically in the animal after dosing and excreted mainly in dung, where insecticidal residues persist which are known to have significant detrimental effects on the invertebrate fauna. The invertebrate fauna of animal dung, particularly cow pats, is an important food source for Lapwing chicks and other birds. The drugs therefore have potential to undermine fledging success in Lapwings, particularly if used on a wide scale during the nesting season (McCracken & Bignal 1991, McCracken 1993, McCracken 1995, Cooke 1997). McCracken (1995) suggested that where such problems could arise the timing and methods of applying avermectins should be considered carefully and a programme of annual variation in the types of anthelmintics used also followed. The latter is, in any case, a sensible policy with all pesticides to avoid the build-up of intractable resistance in target organisms. There is also a need for continued monitoring of the possible environmental effects of using such animal medicines.

One theme running strongly through discussions of the impact of agriculture on Lapwings is the decline in the rates of clutch replacement in the face of agricultural intensification. This seriously erodes productivity, a crucial component of which in Lapwings is an ability to replace lost nests readily. Declines in other farmland birds, for example the Song Thrush (Thomson & Cotton 2000) and the Skylark (Donald 2004), have been similarly linked to a declining ability to replace lost nests or to achieve enough nesting attempts to rear sufficient young, which amounts to the same thing. In Lapwings, furthermore, rates of adult survival have not declined during the period of population decline (Peach *et al.* 1994) nor has clutch size (Chapter 10) nor brood size in successful nests (Shrubb 1990).

From this summary some general principles for Lapwing conservation can be deduced. The species needs nesting grounds with bare ground or short vegetation, feeding sites also with short vegetation, good food supply for adults in the pre-nesting period (which need not be in nesting fields) and for chicks close to nest sites, and preferably some water. Most importantly these conditions need to persist through the nesting season from mid-March to June to allow for adequate levels of clutch replacement. Mixed farming systems of grass and stock, cereals and roots, and low intensity grassland management once created and maintained such conditions throughout farmland.

Sutherland (2004) pointed out that, because much of the British landscape comprises ancient farmland, conservation today is largely aimed at maintaining intervention at a given stage, usually by reintroducing management systems which preserve or replicate low intensity farming systems, through devices such as agri-environment schemes. The point about agri-environment schemes in a managed landscape aimed at species such as the Lapwing is that management must continue or vegetational succession will undermine them. Simply ceasing to farm grassland, for example, achieves little.

One requirement for the re-establishment of Lapwings as widespread and common farmland birds seems to me to be the presence of substantial core populations producing surplus colonists. Such core sites are increasingly being created and managed in grassland nature reserves or, in England, on extensive grassland Environmentally Sensitive Areas (ESAs. Chapter 4). Whilst nature reserves can be straightforwardly managed for conservation, participation in ESAs is voluntary and the management of land joining such schemes has to strike a balance between the needs of agriculture and of conservation, since agriculture remains a significant land-use. A particular difficulty for waders is often a lack of areas of surface flooding, whose maintenance may cause significant difficulties in farm management (Wilson *et al.* 2004). Water levels may also be partly out of the control of site managers because of the impact on water tables of water extraction and drainage in nearby areas. In addition, only part of the land within the designated boundaries of ESAs is covered by agreements.

The main agency creating and managing wet grassland nature reserves in Britain is the RSPB, which in 2005 managed 6,220ha holding 1,450 pairs of Lapwings (E. Andrews pers. comm.). Although densities are high, this is still only a small part (*c.*2%) of the national population. However, these reserves have much greater significance on a regional basis. Thus 3,094ha of reserves in East Anglia held 9–35% of the total regional population found in 1998 (Ausden & Hirons 2002) and, in Wales, the newly created wet grassland area at Ynys-hir already attracts 3–11% of the total Welsh population, which appears unlikely to survive without such refuges.

Much of the management of wet grassland areas revolves around the manipulation of water levels. Most sites need the creation or restoration of a significant infrastructure involving ditch widening and reprofiling, the creation of pools and scrapes, constructing or repairing sluices, dams and culverts, bunding areas and so on, so that water can be held or removed as required (Ausden & Hirons 2002). Annual management particularly concerns grazing and, for Lapwings, involves intensive grazing in winter to create the very short swards needed for nesting in spring. Cattle are preferable for this purpose as their treading breaks up soft rush. Where necessary soft rush is mown off in winter before flooding, which will then control it. Population recovery and increase by Lapwings offered this type of management can be very rapid. At Ynys-hir a struggling population of <20 pairs increased to over 80 pairs in less than five years.

Environmentally Sensitive Areas are part of the Common Agricultural Policy (CAP) of the EU. They are therefore a European measure and they aim to encourage farmers to farm in ways more sensitive to the environment and landscape, in return for payments. Basic options in creating ESAs aim at maintaining the existing landscape by preventing cultivation, limiting reseeding and restricting fertiliser use and the timing of farm operations (Ausden & Hirons 2002). These measures may not be sufficient to maintain numbers of waders. Wilson *et al.* (2005) found that numbers of Lapwings in five key wet grassland areas for waders in England designated as ESAs, the Avon Valley ESA, Norfolk Broads ESA, North Kent Marshes ESA, Somerset Levels and Moors ESA and Suffolk River Valleys ESA, declined by an overall 22% between 1982 and 2002, although numbers increased in the North Kent Marshes. Nevertheless overall decline in these sites was much lower than in undesignated sites and these areas have value in creating more sympathetic hinterlands to areas more directly managed for waders. Higher tier options are available within ESAs, attracting higher levels of payments for enhancing the environment but requiring greater levels of commitment and management from the farmer. Wilson *et al.* (2005) observed that the effectiveness of ESAs as a mechanism for conserving breeding wader populations would be greatly improved by encouraging more landowners to adopt the higher tier options and through the provision of detailed advice on management. Milsom *et al.* (2000) also suggested that conservation prescriptions could be targeted with advantage at the best sites within ESAs, allowing management restrictions to be eased on sites of less quality.

Considerable research effort has been invested in developing suitable techniques for managing wet grassland ESAs to support thriving wader populations, particularly in the management of stocking rates, the timing of grazing, the management of swards by grazing and the manipulation of areas of standing water or flooding (see Milsom *et al.* 2000, Milsom *et al.* 2002, Ausden & Hirons 2002,

Hart *et al.* 2002, Wilson *et al.* 2004). Understanding of how to manage wet grassland for waders is therefore readily available. The ultimate need is certainly to persuade landowners of the value and requirement for investment in such management. Payments need to reflect this.

A second and more widely dispersed type of source population could exist in habitats exploited by Lapwings outside farmland. Such habitats held 8.2% of the total pairs recorded in the 1998 Lapwing survey of England and Wales and numbers had increased in these sites by 6.5% since 1987 (Wilson *et al.* 2001). If the percentage of the population shown is representative of the country-wide occupation of such land, then nearly as many pairs inhabit it (*c.*5,166 pairs) as were found in lowland wet grassland sites by Wilson *et al.* (2005: 5,387 pairs). Such sites are not necessarily subject to less environmental pressure than farmland as many are industrial sites and some are potentially vulnerable to further development. Furthermore, although widely distributed, individual sites may often not be extensive enough to carry a large Lapwing population. Nevertheless, they are worth detailed consideration in Lapwing conservation, as they provide a valuable reservoir of successful breeding birds. Nesting success is often considerably higher than in farmland, at 72% of nests compared to 54% in farmland overall, and clutch replacement rates are better (Figure 10.2) because of the birds' immunity from farming activity (Shrubb 1990). Where sites are controlled by industrial concerns they may be willing to invest in conservation.

One cannot say with certainty that such a policy of creating core habitats with good Lapwing populations will necessarily lead to population recovery more generally. Nevertheless the ecology and history of the species outlined in the preceding chapters seems to me to make it very unlikely that general recovery will take place without such a population engine. The high degree of philopatry exhibited by Lapwings (p.121) is a factor here. Thompson *et al.* (1994) considered that recruitment was mainly driven by philopatry and thus breeding success in a given area. If that declines so do populations. It may thus be rather pointless to invest in habitat creation or Lapwing-friendly agri-environment schemes too far distant from any source population, but a policy of creating core sources of population also needs to be supported by sympathetic management in the general countryside, to encourage more widespread settlement to radiate out from the core.

In the wider countryside a number of options available under agri-environment schemes, such as the Countryside Stewardship Scheme (CSS) in England, Tir Gofal in Wales and the Rural Stewardship Scheme (RSS) in Scotland, offer potential for creating nesting habitat for Lapwings. In England many of these options cluster around spring tillage of various kinds, for example overwintered stubbles followed by cereals or root crops or fallows. A particularly successful option appears to be overwintered stubble followed by spring/summer fallow (Prescription code OS3 under CSS), which appears to have been well supported (Sheldon *et al.* 2004). They noted that it provided nesting habitat throughout the breeding season, attracted higher densities of Lapwings than other crop types and provided brood rearing habitat. Fallowing set-aside would almost certainly achieve similar results and in

Britain a derogation to do this is available from the Department of the Environment, Fisheries and Rural Affairs (DEFRA).

In Wales and Scotland prescriptions for managing grassland tend to be more important. Such prescriptions aim at low stocking rates, limiting fertiliser use, restricting the timing of mowing and other cultivations and usually at taking grazing animals off entirely for the nesting period from mid-March to June.

Not all farmers wish to participate in agri-environment schemes but suitable crops, such as spring cereals, maize, peas and sugar beet remain important in British agriculture and a simpler programme of providing conservation advice and inducement on how best to tweak the management of such crops to birds' advantage may be another way forward (Henderson *et al.* 2004a). One problem with peas and maize may be the timing of cultivation of crops which are more susceptible to frost than cereals, so that sowing times may be pushed into late April, clashing with advanced stages of incubation in Lapwings. Conservation field margins may also be useful as Lapwings breeding on tilled land often take their chicks to such sites for rearing (e.g. Matter 1982, Henderson *et al.* 2004a).

The scale of such schemes is important. Although they do not need to be universal and, for Lapwings, an element of targeting is certainly desirable, a reasonably wide distribution at farm level would do much to restore the overall diversity of farmland. Schemes to benefit Lapwings are likely to benefit many other farmland birds. An important aspect of all such schemes is that they are supported and encouraged by adequate publicity, monitoring and research-based advice. Farmers need to be fully persuaded that changes in farm management are necessary for the recovery of declining bird populations and that they work. The most successful schemes will always be those where their management becomes part of the farm's routine because of the farmer's own interest. Where such schemes are simply regarded as an additional source of income, without input from the farm, they are ineffective.

Sutherland (2004) in fact noted that experience of many agri-environment schemes was that they enjoyed only patchy success. One problem here is that politicians tend to aim for maximum numbers of farmers in the scheme but make no allowance for funding adequate monitoring and advice to follow, which is essential if schemes are to be effective. This wastes money and is ultimately discouraging for farmers. It would be far better to think in terms of fewer agreements, which are more selective and properly monitored and supported. Selection is important to ensure that farms entered for agri-environment schemes aimed at breeding Lapwings have a history of their presence other than just as wintering flocks. This confusion may often arise. One wonders why this is such an alien concept to the political mind. A related political problem is the limited life of many schemes, which are then changed for something different. Some long-term commitment would be preferable.

Many of prescriptions for management in such schemes also seem to be counsels of perfection. Reading them I find myself asking how on earth Lapwings managed in the past for the conditions imposed never existed in farmland in my own experience. Some flexibility may be advisable. For example, in grassland banning grazing from mid-March to June, which is the peak growth period for grass in most lowland areas, risks excessive growth truncating the laying season. In some cases it may be better to graze right through the nesting season at low density to exert some control on growth, so that the birds can replace clutches, which will always be lost, well into May. This is what happened in the past and my experience is that adult cattle at low densities (around one animal/ha) cause few problems for nesting Lapwings. Then, although the scale of grazing is laid down, it would make better sense to define it by density (number of animals per unit area) rather than by rate (livestock units per unit area) because the latter leaves too great a margin of variation: one livestock unit is one dairy cow but seven adult sheep or up to 20 lambs (see Plate 21). Then nothing is said about the type of animal but adult cattle are preferable to young stock and sheep should be avoided. Lapwings appear to have evolved no method of defending their nests from sheep, which not only trample them but may also eat their contents (Chapter 11); sheep also attract crows to nesting fields and probably foxes. Close grazing by sheep in winter, however, may be advantageous in creating swards suitable for nesting. Study of the historical ecology of Lapwings in areas such as the Broadland Marshes would make a useful basis for designing these prescriptions.

There is also the broader issue of landscape-scale conservation to provide a sympathetic background to the more detailed work in reserves and agri-environment schemes. Sutherland (2004) drew attention to a number of benefits that could be derived from the landscape, of which three seem to be particularly relevant in this discussion: flood reduction, improving water quality and supply and tourism.

Tourism, now a more significant activity in the UK economy than farming, is becoming increasingly important in many areas, especially those unsuitable for intensive agriculture. It benefits from more diverse landscapes. So does wildlife. Sutherland quoted the example of maintaining flower-rich hay meadows for their beauty and attraction. The maintenance of such semi-natural habitats may be a more successful and economic form of management in areas such as national parks than subsidising agriculture there.

Flood reduction would be greatly helped by restoring the catchments of many rivers, particularly in the uplands. Upland catchments once acted as sponges, holding and storing their high rainfall and releasing it more slowly into rivers and streams. The impact of drainage and intensive grazing by high stocking rates of sheep has been to alter these characteristics fundamentally. Intensive grazing by sheep greatly reduces vegetation cover and the trampling of sheep compacts the soil. Both significantly increase erosion and run-off, promoting damaging flooding downstream in river valleys. Restoring the hydrology of upland catchments to limit such flooding would be far more economic than expensive flood prevention engineering works. It would also have significant gains for wildlife, restoring important wetland habitats for waders. This may become of increasing value if climate change has the predicted impact on sea levels, inundating many lowland wetlands.

Sutherland (2004) also pointed out that, in a future likely to need to husband water resources more carefully, maintaining water tables and protecting and restoring water catchments could become major issues in land management. Species such as Lapwings could only benefit from such a policy.

One final point to consider is the status of wintering populations, although there is probably little reason to regard these as of conservation concern at present. Nevertheless conservation interests might give some thought to explaining clearly how it is that British farmland continues to attract a large wintering population whilst it can only hold a declining breeding one. At present most Lapwings in winter are found in arable land but this represents a quite recent change in habits. Any discussion of the conservation of winter populations must include two points made by Gillings (2003): both the diurnal and nocturnal patterns of behaviour by the species must be considered, as must also variations in distribution and habitat use related to variations in winter climate (Chapter 6). The winter distribution today differs sharply from that 20 years ago (Chapter 6). In the future certain changes, proposed or in place, may limit the attractiveness of arable land. One is the change in the timing of sowing winter cereals. September sowing is now common and may well restrict the attractiveness of cereals to plovers because of its impact on growth patterns (p.87). Another is the proposed changes in sugar beet growing which may significantly limit the area grown. Harvested sugar beet fields are favoured winter feeding sites and the crop's presence in arable rotations encourages some later sowing of winter cereals (p.87). Whether such changes would have any impact on overall wintering populations, which have the option of shifting back into grassland habitats, is unclear. Such a shift would, in any case, presumably occur if the winter climate reverts to the colder regimes of the recent past. The protection of the present areas of old pasture in the Midlands would therefore be a sensible precaution for the Lapwing population in the future.

APPENDIX 1

Changes in the breeding populations and ranges of Lapwings in Europe in the 19th and 20th centuries

Austria. Glutz von Blotzheim *et al.* (1975) estimated 1,500 pairs, which have increased to 3,000–6,000 (Birdlife International 2004).

Belarus. Perhaps fairly stable. Reported as declining in Birdlife International (2004) but as increasing in Thorup (2005).

Belgium. Marked increase and extension of range into southern Belgium in last 40–50 years with exploitation of arable land following loss of traditional grassland habitats; 32% of population now breeds in southern Belgium. Devillers *et al.* (1988) considered that there were 15,000–20,000 pairs and the present estimate of 17,000–24,000 (Thorup 2005) suggests some continued increase.

Britain. 19th century accounts show that there was some range expansion and a clear increase in population on the northern Scottish mainland from about the 1840s (Harvie-Brown & Buckley 1895, Harvie-Brown & MacPherson 1904) and the population probably continued to increase there into the 1950s (Baxter & Rintoul 1953, Spencer 1953). In the northern islands Lapwings were rare in Orkney in the early 19th century but increasing and abundant by mid-century, although declining with excessive egging at its end (Buckley & Harvie-Brown 1891). In Shetland absent as a breeding bird in the late 18th century but Saxby (in Venables & Venables 1955) noted it as a very rare breeder in the 19th century until a colony established itself on the south side of Baltasound, Unst, between 1854 and 1858. By 1871 quite common throughout Shetland and increase continued well into the first half of the 20th century (Venables & Venables 1955). In the Outer Hebrides the species has long been at least locally numerous but various authorities (summarised in Spencer 1953) noted the population increasing into the 1950s.

Otherwise county and regional avifaunas provide evidence of long-term decline, at least regionally, from the 19th century to the present, although there have been significant fluctuations. Figures published by Shrubb & Lack (1991) suggested a 61% decline in England and Wales between 1956–65 and 1987, when they estimated a population of 123,134 pairs there. Further sharp decline, by 49%, shown by Wilson *et al.* (2001) to *c.*63,000 pairs in 1998. Decline greatest in Wales, where Lovegrove *et al.* (1995) estimated at least 15,000 pairs in 1970, which had fallen to 1,689 in 1998, a decline of 89%. In Scotland population more stable. Thom (1986) estimated 75,000–100,000 pairs and O'Brien (1996) estimated 92,000 pairs in lowland Scotland (but see discussion on p.47). Some decline since and O'Brien *et al.* (2002) and O'Brien & White (2003) estimated 91,505 pairs for the whole of Scotland in 1997–2000.

Bulgaria. Apparently stable but breeding densities vary on species' southern boundary (Nankinov 1989). Some recent decline possible and 600–1,000 pairs estimated in 1997–2002 (Birdlife International 2004).

Czechoslovakia. Hromadkova 1987 noted a decline which has continued. 7,000–10,000 pairs in 2000, a decline of 50–79% (Birdlife International 2004).

Denmark. Unlike other Scandinavian countries has declined for much of 20th century with habitat change in agriculture, particularly drainage of wet grasslands. Farmland populations have declined by 80% in the past 25 years and, overall, *c.*100,000 pairs in 1976–1980 have declined to 30,000–45,000. (Frikke 1991, Thorup 2005).

Estonia. Population increase from second half of 19th century. The present estimate of 15,000–30,000 pairs is considerably higher than the 10,000 pairs estimated in the 1970s, although declines were noted in all habitats in 1994. (Leibak *et al.* 1994, Thorup 2005).

Finland. Marked increase and spread from late 19th century, during which has colonised whole country north to Lake Inari. Northward spread particularly rapid after 1940; 26,000 pairs estimated in 1950s increasing to 100,000–200,000 in early 1980s. Sharp decline since and 70,000–100,000 pairs in 1992 and 50,000–80,000 in 1998–2002. Merikallio 1958, von Haartman 1973, Koskimies 1989, 1992, Birdlife International 2004

France. Not common in 19th century. Rapid increase in numbers and range in first half of 20th, particularly from 1940s, extending into 1970s. Estimated maximum population of 39,500–40,000 pairs in 1961; numbers halved to 21,000 in 1979 and still decreasing. Surveys in 1984 and 1995–96 recorded 17,400–20,300 and 12,716–16,073 (since revised to 18,000) pairs respectively. Yeatman 1976, Dubois *et al.* 1991, Deceuninck & Mahéo 1998. Birdlife International (2004) put population at 17,000–20,000 pairs during 1998–2002.

Germany. In many areas marked decrease due to habitat change and egging in late 19th century and into 20th. Nadir reached 1920s and 1930s. Increase from 1940s, with adaptation to breeding in arable. 215,000 pairs estimated in 1980. But sharp decline, of 63%, noted Schleswig Holstein 1970–92, with drainage of wet meadows and steeper declines, of up to 90%, noted in some areas, as birds attracted to nesting in forage maize crops, a new habitat. National population of 67,000–104,000 estimated 1996–1999. Glutz von Blotzheim *et al.* 1975, Rheinwald 1993, Klinner 1991, Busche 1994, Birdlife International 2004.

Hungary. Gorman (1996) noted that population probably stable but Birdlife International (2004) noted some decline 1999–2002.

Ireland. Figures given by Hötker (1991) and Madden *et al.* (1998) indicate an overall decline of 36% since the mid-1980s. In Ulster the decline was 66% between 1987 and 1999, to 1,771 pairs. (Henderson *et al.* 2002).

Italy. Only a few pairs bred before 1950. Great increase in 1970s and 700–1,000 pairs estimated 1980–85 and 1,700–1,900 in 2003. Increase considered due to decline in persecution. (Tinarelli & Bacetti 1989, Birdlife International 2004).

Luxembourg. Breeding started in 1960 after an absence of *c.*30 years and numbers of breeding pairs increased from 100 in 1967 to over 200 in 1973: decline to only 20–30 since. (Glutz von Blotzheim *et al.* 1975, Birdlife International 2004).

The Netherlands. Klomp (1954) considered marked decline over previous 30 years, with grassland improvement, particularly reseeding and increased fertiliser usage. Increase apparent since. 120,000 pairs estimated in 1970s, 125,000–165,000 in 1982, 226,000–278,000 in 1989, 200,000–300,000 in 1998/2000. Beintema noted that increases partly arose from better survey coverage and efficiency but unlikely that scale indicated stems only from such a source. Population now considered stable 1998–2000 with a slow decline in grassland balanced by an increase on arable land. (Klomp 1954, Rooth 1977, Teixeira 1979, Cramp & Simmons 1982, Beintema 1991, SOVON 2000, Birdlife International 2004).

Norway. Marked increase and spread from late 19th century. Confined to the coast north to 66°N in the 1890s but has since expanded to breed virtually throughout the country, penetrating east Norway from about 1920. Population peaked at *c.*60,000 pairs in early 1980s and and broadly stable at 40,000–80,000 pairs thereafter. (Haftorn 1958, Gjershaug *et al.* 1994, Thorup 2005).

Poland. During early 20th century spread into 'fields', implying a spread into arable land. Use of such habitat was declining in the 1970s and a more general decrease in the west emerging in the 1980s. (Glutz von Blotzheim *et al.* 1975, Tomialojc 1987.)

Russia and Siberia. The northward expansion of range in Russia and Siberia was fully summarised by Tomkovich (1992). Essentially the expansion started in the Arkhangel'sk region as early as 1939 but it has been most marked along the whole northern boundary of the species' range from the 1950s. The expansion has tended to follow the major river valleys: the Pechora, Ob, Yenisey, Lena and Amur. It has been least marked in the Yenisey valley, which Tomkovich remarked as little developed agriculturally. The isolated Yakutiya population was first discovered along the middle Vilyuy river (a tributary of the Lena) in 1967 and it has increased rapidly since. In the 1980s it may have merged with the main range to the south but this is not definitely known. The species now breeds north to the Murmansk region in the extreme west, to 65°30′N on the Pechora, to about 67°N on the Ob, to 60°30′N on the Yenisey, to about 64°N in Yakutiya and 54°N on the Sea of Okhotsk and the Amur. Birdlife International (2004) considered that population declined by 20–29% during 1990–2000.

Spain. Cramp & Simmons (1982) suggested some increase in north and east, which comparison of Voous (1960) and Purroy (1997) also suggests. The 2003 Breeding Bird Atlas, covering 1998–2002, found some 1,600 pairs, 80% of them in the central plains of Castilla-La Mancha and Castilla y León. Population trend stable but numbers fluctuate considerably between years according to rainfall (Purroy 1997, Martí & Del Moral 2003, Thorup 2005).

Sweden. As in Norway and Finland a marked expansion of range and population in 20th century. In early 1970s Ulfstrand & Högstedt (1976) estimated 120,000 pairs and Koskimies (1992) put population at 100,000–150,000 pairs and noted marked decline, which has continued to the present level of 50,000–100,000 pairs (Birdlife International 2004).

Switzerland. Increase from 130 pairs in 1930s to *c.*1,000 in early 1970s. Decline since, perhaps because of poor breeding success and 450 pairs in 1998 and 250–400 in 1998–2002. H. Matter in Schifferli *et al.* 1980, Schmid *et al.* 1998, Birdlife International 2004.

Ukraine. Declines reported in seven regions in the north and west but some increase in the Crimea (in Thorup 2005).

APPENDIX 2

Habitats used by breeding Lapwings in Europe

References are as Appendix 1 with additions.

Belgium. Originally a breeding bird of grasslands, particularly old wet pastures in the north. With loss in area of such habitats to drainage, improvement and arable farming and to urban development, species has colonised arable land extensively, expanding its range into southern Belgium. Breeding productivity in arable land reported to be poor. (Devos *et al.* 1991).

Britain. Historically particularly a species of the Waste (see p.44). With enclosure and the development of rotation farming, adapted to mixed farmland but wet grasslands always important. See Chapter 4 for a detailed discussion, particularly Tables 4.2 and 4.3.

Czechoslovakia. Now beginning to spread into dry agricultural land, following loss of wet grassland.

Denmark. Traditionally a wet grassland species, including wet saltmarshes. With extensive loss and conversion of this habitat to arable land species became more abundant in arable land in some regions. Nevertheless a strongly significant correlation found between size and density of Lapwing populations and area of permanent grassland in Tøndermasken and Ettrup & Bak considered that the population is mainly supported by high level of successful breeding in wet saltmarshes. Productivity in arable land considered below level for self-maintenance and steep decline in farmland populations since 1970s ascribed mainly to the large scale loss of spring tillage. (Ettrup & Bak 1985, Gram *et al.* 1990 in Frikke 1991).

Estonia. In 19th century inhabited wet meadows, open fens and transition bogs, spreading in second half of century into drained meadows and during first half of 20th century into dry meadows and arable fields.

France. Most breed on grasslands, preferably moist pastures or meadows. Yeatman (1976), for example, showed important concentrations in the marshlands of western France, where there were some 200,000ha of wet grassland between Nantes and Bordeaux, now being extensively drained and converted to arable land (Leroux 1991). Arable land now second most important nesting habitat and, in some Departments, the most important. Dry and improved grassland little used. Productivity on arable land found insufficient for self-maintenance.

Finland. Population increase in agricultural land associated with extensive under-drainage to replace open ditches, removal of trees and scrub, decline in the area of leys, natural meadows and pastures and increase in area of cereals and peas. With increased numbers breeding birds spread into smaller fields and also open peatlands, which are its main nesting habitat in Lapland, and islets from the early 1950s. Declines followed abandoning of grazing in shore (salt) marshes as vegetation grew. Productivity on intensively managed farmland considered too low for self-maintenance of population. Population decline since late 1970s coincided with extensive abandonment of agricultural land (total farm holdings fell from 400,000 to 255,000 from 1969 to 1975) and switch to farm forestry from crops and stock. (Hildén 1965, Varjo 1984, Solonen 1985, Tiainen *et al.* 1985, Ratcliffe 2005).

Germany. Extensively farmed old meadows and pastures and, in the north especially, wet grassland. From the 1940s adapted to arable habitats. More recently loss of wet grassland has led to increasing use of forage maize, often grown in wet fields which leads to main period of cultivation clashing with incubation and hatching, leading to very high losses from both ploughing and spreading slurry. Increasingly in these areas pairs do not lay repeat clutches and there is a rising trend for birds to summer without nesting.

Hungary. Dry grassy and saline *puszta,* edges of salt lakes, pastures, meadows, ploughed fields, crops such as maize, lucerne and cereals and, in the uplands, stream valleys and high meadows.

Ireland. In 19th century bred mainly on flat moors and marshy lands, islands on inland lakes and sandy tracts on coast; no mention made of improved farmland (Ussher & Warren 1900). Kennedy *et al.* (1954) described same range of habitats but also noted 'sometimes meadows and pastures', whilst Ruttledge (1966) added occasional nesting on arable. Partridge & Smith (1992) for Ulster gave 68% of population nesting in damp grassland, 17% on raised bog, 4% on blanket bog, 2% on fen, 5% on waste ground and 4% on arable land, mainly spring cereals. Henderson *et al.* (2002) found marsh/fen, unimproved grassland and arable land selected; bog/mire, heather moor and improved grassland avoided. Densities of 8–10 pairs/km^2 occur in major wetlands (Nairn *et al.* 1988).

Italy. Maize fields now commonest nesting habitat, followed by meadows and wet grassland.

Latvia. Cultivated meadows, pastures, arable land, occasionally waste land, meadows on lakes and rivers and by the sea, bogs, mainly transitional and high bogs. (Viksnes 1989).

The Netherlands. Damp grasslands, although entirely artificial. Highest densities occur in lowest areas. Klomp (1954) identified three factors influencing nesting habitat selection—absence of trees, grey-brown or grey-green colour to the grass indicating slow growth and low vegetation or bare ground. Well distributed on higher ground, where 40–80% nest on arable land, particularly forage maize.

Poland. Mainly a marsh and wet grassland species, with important concentrations, for example, in the Notec Valley, Biebrza Marshes and lower Bug Valley. Some tendency to colonise 'fields' (arable land?) in early 20th century. In south Poland drained ponds, wet meadows, and, less, often dry meadows and fields in river valleys near ponds. (Gromadzka *et al.* 1985, Walasz & Mielczarek 1992).

Russia and Siberia. Most frequently damp grassy meadows and marshes, stubble fields and winter crops. In drier areas often subsidences on the steppe. In European Russia becoming a bird of fields rather than marshes (Dementiev & Gladkov 1969). Range expansion throughout the northern part of range exclusively in extensive 'water meadows' (presumably flood meadows or wet grassland) in the major river valleys, mainly man-made for stock grazing and hay crops. There was a much less marked expansion in the Yenisey valley where only limited agricultural expansion. With geographic expansion then spread into natural habitats, saltmarshes on the White Sea and the Sea of Okhotsk and natural steppe grasslands in Yakutiya.

Spain. Breeds in colonies in wet meadows, crops and fallows; often near water and generally in seasonally flooded areas below 1,000m. Most widely distributed in wet years. (Martí & Del Moral 2003).

Sweden. Saltmarsh pastures, fresh pastures and spring tillage; peatlands in the north. From the 1950s spread onto islets in the outer archipelago. In central Sweden a strong preference for nesting on tilled land, usually close to flooded tillage and grassland. Since 1950 increased intensity of farming has led to large scale change in the agricultural landscape. Meadows and pastures have declined drastically in area, saltmarsh pastures have been abandoned and large scale cereal fields lacking ditches or edge vegetation have been created. Recently extensive areas of set-aside fallows, amounting to 12.5% of farmland by 1992, have been created. Much old grassland has been lost and there is a shortage of grazing animals in many areas. (Hildén 1965, Larsson 1969, Robertson & Berg 1992, Berg *et al.* 1992, Berg 1993, Berg & Part 1994, Berg 2002. Ratcliffe 2005).

Switzerland. Originally occupied only marshlands and meadows and population fell to very low level as these were replaced by arable land. Species than adapted to arable habitats particularly following the opening up of farmland habitats by hedge removal and expansion of maize growing. 85% breed on wettish soils on the silted areas of post glacial lakes. Breeding success considered too poor in arable habitats to maintain population, which thought dependent on immigration from elsewhere. (Imboden 1970, 1971, Matter 1982)

APPENDIX 3

The diet of the Lapwing

Britain. In the mid-19th century, wireworms (larval click-beetles, Elateridae) were widely recorded as important prey and Yarrell (1845) also recorded earthworms (Lumbricidae), slugs and beetles. Archibald (1908) listed slugs, caterpillars, leatherjackets (larval crane flies, Tipulidae), wireworms, earwigs, earthworms, beetles and caddisflies as prey, with the first four most important. Harvie-Brown (1906) quoted H. Drummond Hay that, with the sharp decline of Lapwings in the Tay region in the 19th century, there was a marked increase of slug problems in winter wheat. Collinge (1924–27) found 19 weevils (Curculionidae), 23 unidentified larvae, 11 slugs, 53 wireworms, 47 leatherjackets, 22 crane flies and 43 beetles in three adults collected in June and July (no localities). 69 stomachs (no localities or dates) contained earthworms (two species), woodlice, millipedes (two species), earwigs, Hemiptera (one species), beetles (17 species and unidentified larvae), caddis flies, Lepidoptera (six species, presumably as larvae), other Diptera (two species), Hymenoptera (larva of one species), slugs and snails (five species). Also seeds of *Ranunculus, Spergula* and *Polygonum.* No quantities given. In Cumbria E. Blezard found one bird in April to contain earthworms, a wireworm and insects; one in September held noctuid moth larvae, weevils, wireworms and grass; two in October had moth larvae, a beetle and grass (Cramp & Simmons 1982).

In the North Pennines, nine stomachs collected before 1 May contained nine whole earthworms, 22 adult and 17 larval beetles and 50 dipteran larvae; seven collected after 30 April (included four from chicks) contained 80 adult and four larval beetles, two adult and ten larval Diptera and five other items (Baines 1990). In northern England breeding adults fed largely on earthworms, chicks principally on beetles (Baines 1994). In Stirlingshire Galbraith (1989a,b) also found that earthworms and leatherjackets were important prey for adults in the early breeding season. They remained so for chicks. Of 168 items found in 10 Lapwing chicks in arable land there 46.7% by number and 14.5% by dry weight were beetles, 19% and 22.4% tipulid larvae, 9.5% and 3.4% spiders, 4.8% and 53.1% earthworms, and 19.6% and 6.4% other items. Of 223 items in ten chicks in rough grazing there, 69.6% by number and 27.4% by dry weight were beetles, 6.2% and 9% tipulid larvae, 10.6% and 4.7% spiders, 3.9% and 55.2% earthworms and 8.9% and 3.2% others. In coastal grazing marsh in Kent adult diets in April comprised 35% earthworms, 45% tipulid larvae, 5% lepidopteran larvae and 13% beetles; in May, earthworms declined to 17%, lepidopteran larvae increased to 13%, tipulid larvae comprised 36% and beetles 28%; in June earthworms comprised 11%, lepidopteran larvae 12%, dipteran larvae 58% and beetles 21%. Chicks on the same site took 82% dipteran larvae, mainly chironomids (64%), and 17% beetles in one year, but in the next 19% earthworms and 25% beetles, with smaller percentages of larval Diptera (Ausden *et al.* 2003).

In winter largely earthworms in pasture (e.g. Barnard & Thompson 1985, Tucker 1992, Village & Westwood 1994). In arable land 3–15% earthworms and remainder other items, mainly surface invertebrates (Shrubb 1988, Gillings 2003).

Finland. 12 birds contained five earthworms and a lepidopteran larva after cold spring; beforehand one contained only Diptera (Vepsäläinen 1968).

France. Analysis of 54 stomachs from Northern France throughout the year (22 for August) revealed weevils in 72%, Scarabeidae in 15%, click-beetles in 16%, gastropods in 20%, Lepidoptera in 11% and traces of vegetation in 70% (P. Madon in Cramp & Simmons 1982).

Germany. In North Germany 19 stomachs of chicks 2–8 days old contained 207 beetles, 90 insect larvae, 12 Diptera and single myriapod and earthworm (Matter 1982).

Netherlands. Earthworms normally staple diet in summer (Klomp 1953). But in dry summer of 1959 56 stomachs analysed by Voous (1962) held no earthworms but 52% had noctuid larvae, 47% earwigs, 25% ground beetles, 29% weevils and 60% spiders. Analysis of the faeces of 262 chicks found earthworms in 81%, gastropods in 24%, Nematocera in 30%, tipulids in 10%, Chironomidae in 13%, Brachycera and Cyclorrhapha in 49%, Hymenoptera in 21%, beetles in 92%, spiders in 10%; larvae of tipulids were found in 37%, Stratiomyidae (49%), other Diptera (26%) and beetles (24%) together with small numbers of other invertebrates; seeds were found in 9% (Beintema *et al.* 1991).

Soviet Union. For 19th century Dementiev & Gladkov (1969) gave acridid orthopterans, earthworms, slugs and larvae. In 20th century, on steppes particularly, Acrididae and nocturnal tenebrionid beetles. Ten stomachs from Naurzum State Preserve principally carabid and curculionid beetles, often nocturnal, and larvae of dytiscid diving beetles. During 1945–63 15 stomachs and 1962–63 5 stomachs all contained beetles—Carabidae 47% and 20% by frequency, tenebrionids 20% and 60%, weevils 60% and 80%—with smaller numbers of Dytiscidae, ants, earwigs, Acrididae, cicadas, *Planorbis* snails and some vegetation (V .F. Ryabov & N. I. Mosalova in Cramp & Simmons 1982). In Ukraine, beetles were also very important in 111 stomachs (no dates), with 326 weevils in 61%, 314 carabids in 61%, 736 scarabs in 49% and 237 click-beetles in 9%. Dytiscidae, long-horned beetles (Cerambycidae), pill beetles (Byrrhidae) and water scavengers (Hydrophilidae) also recorded in <6%. Acridids recorded in over 44%, ants in 32%, Diptera in over 11% and smaller numbers of earthworms, molluscs, spiders, mayflies, fish and frogs (O.B. Kistyakivski in Cramp & Simmons 1982).

Sweden. Högstedt (1974) and Blomqvist & Johansson (1995) found that earthworms most important diet for females in pre-laying period. Of 352 prey items recorded from chick faeces in fresh pasture 42.6% crustaceans, 34.7% Diptera, mainly tipulids, and 18.5% beetles of 14 families; some insects and spiders also taken. Of 256 items similarly collected on shore pastures 39.5% rag-worms (Nereidae), 43% crustaceans and 10.9% beetles of eight families, together with a small percentage of other insects (Johansson & Blomqvist 1996).

Switzerland. 20 stomachs of chicks 2–8 days old contained 24 beetles, 61 insect larvae, three Diptera, eight myriapods and seven earthworms (Matter 1982).

APPENDIX 4

Scientific names of species mentioned in the text

PLANTS	
Cranberry	*Vaccinium oxycoccus*
Yellow Flag	*Iris pseudacorus*
Soft Rush	*Juncus effusus*
Purple Moor Grass	*Molinia caerulea*
INVERTEBRATES (see also Appendix 3)	
Turnip Moth	*Agrotis segetum*
Turnip Sawfly	*Athalea rosae*
MAMMALS	
Badger	*Meles meles*
Buffalo	*Syncerus caffer*
Elephant	*Loxodonta africana*
Hare	*Lepus europaeus*
Hedgehog	*Erinaceus europaeus*
Hippopotamus	*Hippopotamus amphibius*
Mink	*Mustela vison*
Polecat	*Mustela putoris*
Rabbit	*Oryctolagus cuniculus*
Red Fox	*Vulpes vulpes*
Stoat	*Mustela erminea*
Warthog	*Phacochoerus africanus*
BIRDS	
Andean Lapwing	*Vanellus resplendens*
Arctic Tern	*Sterna paradisaea*
Avocet	*Recurvirostra avosetta*
Banded Lapwing	*Vanellus tricolor*
Black-headed Gull	*Larus ridibundus*
Black-headed Lapwing	*Vanellus tectus*
Black Kite	*Milvus migrans*
Blacksmith Lapwing	*Vanellus armatus*
Black-tailed Godwit	*Limosa limosa*
Black-winged Lapwing	*Vanellus melanopterus*
Brown-chested Lapwing	*Vanellus superciliosus*
Brünnich's Guillemot	*Uria lomvia*
Canada Goose	*Branta canadensis*
Carrion Crow	*Corvus corone*
Common Buzzard	*Buteo buteo*

Common Guillemot	*Uria aalge*
Common Gull	*Larus canus*
Common Kestrel	*Falco tinnunculus*
Crowned Lapwing	*Vanellus coronatus*
Curlew	*Numenius arquata*
Dunlin	*Calidris alpina*
Eider	*Somateria mollisima*
Gannet	*Morus bassanus*
Golden Plover	*Pluvialis apricaria*
Green Sandpiper	*Tringa ochropus*
Grey-headed Lapwing	*Vanellus cinereus*
Grey Heron	*Ardea cinerea*
Grey Partridge	*Perdix perdix*
Hen Harrier	*Circus cyaneus*
Herring Gull	*Larus argentatus*
Javanese Wattled Lapwing	*Vanellus macropterus*
Killdeer	*Charadrius vociferus*
Kittiwake	*Rissa tridactyla*
Little Ringed Plover	*Charadrius dubius*
Long-toed Lapwing	*Vanellus crassirostris*
Magpie	*Pica pica*
Marsh Harrier	*Circus aeruginosus*
Masked Lapwing	*Vanellus miles*
Meadow Pipit	*Anthus pratensis*
Merlin	*Falco columbarius*
Moorhen	*Gallinula chloropus*
Northern Lapwing	*Vanellus vanellus*
Oystercatcher	*Himantopus ostralegus*
Peregrine	*Falco peregrinus*
Pheasant	*Phasianus colchicus*
Pied Lapwing	*Vanellus cayanus*
Raven	*Corvus corax*
Red Grouse	*Lagopus lagopus*
Red-legged Partridge	*Alectoris rufa*
Redshank	*Tringa totanus*
Red-wattled Lapwing	*Vanellus indicus*
Ringed Plover	*Charadrius hiaticula*
River Lapwing	*Vanellus duvaucelii*
Rook	*Corvus frugilegus*
Ruff	*Philomachus pugnax*
Senegal Lapwing	*Vanellus lugubris*
Senegal Wattled Lapwing	*Vanellus senegallus*
Short-eared Owl	*Asio flammeus*
Skylark	*Alauda arvensis*
Snipe	*Gallinago gallinago*
Sociable Lapwing	*Vanellus gregarius*
Song Thrush	*Turdus philomelos*
Southern Lapwing	*Vanellus chilensis*
Sparrowhawk	*Accipter nisus*
Spot-breasted Lapwing	*Vanellus melanocephelus*
Spur-winged Lapwing	*Vanellus spinosus*
Starling	*Sturnus vulgaris*
Whimbrel	*Numenius phaeopus*
White-headed Lapwing	*Vanellus albiceps*
White-tailed Lapwing	*Vanellus leucurus*
Yellow Wagtail	*Motacilla flava*
Yellow-wattled Lapwing	*Vanellus malabaricus*

References

ADAMIAN, M.S. & KLEM, D. 1997. *Field Guide to the Birds of Armenia.* American University of Armenia.

ALI, S. & RIPLEY, S.D. 1969. *Handbook of the Birds of India and Pakistan.* University Press. Oxford.

AMMONDS, T.R. 1972. The Decline of the Lapwing. *Herefordshire Ornithological Report 1972.*

ANDREWS, I.J. 1995. *Birds of the Hashemite Kingdom of Jordan.* Privately published.

APLIN, O.V. 1889. *The Birds of Oxfordshire.* Oxford University Press.

APPLETON, G. 2002. *Lapwing* in Wernham, C., Toms, M., Marchant, J., Clark, J., Siriwardena, G. & Baillie, S. (eds). *The Migration Atlas: movements of the birds of Britain and Ireland.* T & AD Poyser. London.

APPLETON, G.F. & MINTON, C.D.T. 1978. The primary moult of the Lapwing. *Bird Study* 25: 253–256.

ARCHIBALD, C.F. 1908. Wild birds useful and injurious. Part III. *Journal of the Royal Agricultural Society* lxviii: 1–16.

ASENSIO, B. 1992. Migración e invernada de la Avefría (*Vanellus vanellus*) en la península Ibérica. *Doñana Acta Vertebrata* 19: 71–84.

ASH, J. 1948. Evidence of young-carrying by Lapwing. *British Birds* 41: 356.

ATKINSON, P.W., BUCKINGHAM, D. & MORRIS, A.J. 2004. What factors determine where invertebrate-feeding birds forage in dry agricultural grasslands? *Ibis* 146 (Suppl.2): 99–107.

ATKINSON, P.W., FULLER, R.J., VICKERY, J.A., CONWAY, G.J., TALLOWIN, J.R.B., SMITH, R.E.N., HAYSOM, K.A., INGS, T.C., ASTERAK, E.J. & BROWN, V.K. 2005. Influence of agricultural management, sward structure and food resources on grassland field use by birds in lowland England. *Journal of Applied Ecology* 42: 932–942.

AUSDEN, M. & HIRONS, G.J.M. 2002. Grassland nature reserves for breeding wading birds in England and the implications for the ESA agri-environment scheme. *Biological Conservation* 106: 279–291.

AUSDEN, M., ROWLANDS, A., SUTHERLAND, W.J. & JAMES, R. 2003. Diet of breeding Lapwing *Vanellus vanellus* and Redshank *Tringa totanus* on coastal grazing marsh and implications for habitat management. *Bird Study* 50: 285–293.

AUSTIN, G.E. & REHFISCH, M.M. 2005. Shifting non-breeding distributions of migratory fauna in relation to climate change. *Global Change Biology* 11: 31–38.

BAGG, A.M. 1967. Factors affecting the occurrence of the Eurasian Lapwing in eastern North America. *Living Bird* 6: 87–121.

BAINES, D. 1988. The effects of improvement of upland marginal grasslands on the distribution and density of breeding wading birds (Charadriiformes) in northern England. *Biological Conservation* 45: 221–236.

BAINES, D. 1989. The effects of improvement of upland, marginal grasslands on the breeding success of Lapwings *Vanellus vanellus* and other waders. *Ibis* 131: 497–506.

BAINES, D. 1990. The roles of predation, food and agricultural practice in determining the breeding success of the Lapwing *Vanellus vanellus* on upland grasslands. *Journal of Animal Ecology* 59: 915–929.

BAINES, D. 1994. Factors determining the breeding success and distribution of Lapwings *Vanellus vanellus* on marginal farmland in northern England. In: *The Ecology and Conservation of Lapwings Vanellus vanellus.* UK Nature Conservation No.9. Joint Nature Conservation Committee. Peterborough.

Bak, B. & Ettrup, H. 1982. Studies on the migration and mortality of the Lapwing (*Vanellus vanellus*) in Denmark. *Danish Review of Game Biology* 12: 1–20.

Balança, G. 1984. Migrations et hivernage du Vanneau huppe (*Vanellus vanellus*) et du Pluvier doré (*Pluvialis apricaria*) dans le sud de la Brie: determinisme meteorologique, selection de l'habitat et activités. *L'Oiseaux* 54: 337–349.

Bamford, R. 2002. Dyfi Lapwing research. *Ceredigion Bird Report 2002.* Wildlife Trust of South & West Wales. Cardigan.

Bannerman, D.A. 1961. *The Birds of the British Islands vol x.* Oliver & Boyd. Edinburgh.

Bannerman, D.A. & Bannerman, W.M. 1963, 1966, 1967. *Birds of the Atlantic Islands, vol. 1, 2, & 3.* Oliver & Boyd. Edinburgh.

Bannerman, D.A. & Bannerman, W.M. 1971. *Handbook of the Birds of Cyprus and migrants of the Middle East.* Oliver & Boyd. Edinburgh.

Bannerman, D.A. & Bannerman, W.M. 1983. *The Birds of the Balearics.* Croom Helm. London.

Barnard, C.J. & Thompson, D.B.A. 1985. *Gulls and plovers: the ecology and behaviour of mixed-species feeding groups.* Croom Helm. London.

Barnett, P.R., Whittingham, M.J., Bradbury, R.B. & Wilson J.D. 2004. Use of unimproved and improved lowland grassland by wintering birds in the UK. *Agriculture Ecosystems & Environment* 102: 49–60.

Baxter, E.V. & Rintoul, L.J. 1953. *The Birds of Scotland.* Oliver & Boyd. Edinburgh.

Beaumont, D., Prescott, T., Rout, F., Street, L. & Fraser, P. 1998. The first twenty-five years of the Insh Marshes RSPB Nature Reserve, Inverness-shire. *RSPB Conservation Review No.12* 102–110.

Beintema, A.J. 1981. Meadow birds in the Netherlands. *Research Institute for Nature Management Annual Report.*

Beintema, A.J. 1986. Man-made polders in the Netherlands: a traditional habitat for shorebirds. *Colonial Waterbirds* 9: 196–202.

Beintema, A.J. 1988. Conservation of grassland bird communities in the Netherlands. *ICBP Technical Publication No.7* 105–111.

Beintema, A.J. 1991. Status and conservation of meadow birds in the Netherlands. *Wader Study Group Bulletin* 61, *Supplement*: 12–13.

Beintema, A.J. 1994. Condition indices for wader chicks derived from body-weight and bill-length. *Bird Study* 41: 68–75.

Beintema, A.J. 1995. Fledging success of wader chicks, estimated from ringing data. *Ringing & Migration* 16: 129–139.

Beintema, A.J., Beintema-Hietbrink, R.J. & Müskens, G.J.D.M. 1985. A shift in the timing of breeding in meadow birds. *Ardea* 73: 83–89.

Beintema, A.J. & Müskens, G.J.D.M. 1987. Nesting success of birds breeding in Dutch agricultural grasslands. *Journal of Applied Ecology* 24: 743–758.

Beintema, A.J. & Visser, G.H. 1989a. Growth parameters in chicks of charadriiform birds. *Ardea* 77: 169–180.

Beintema, A.J. & Visser, G.H. 1989b. The effect of weather on time budgets and development of chicks of meadow birds. *Ardea* 77: 181–192.

Beintema, A.J., Thissen, J.B., Tensen, D. & Visser, G.H. 1991. Feeding ecology of charadriiform chicks in agricultural grassland. *Ardea* 79: 31–44.

Beintema, A.J., Dunn, E. & Stroud, D.A. 1997. Birds and wet grasslands. In: Pain, D.J. & Pienkowski, M.W. (eds). *Farming and Birds in Europe.* Academic Press. London.

Belding, E.T. 1971. Drainage survey in England and Wales. *Agriculture* 78: 250–254.

Berg, Å. 1993. Habitat selection by monogamous and polygamous Lapwings on farmland—the importance of foraging habitats and suitable nest sites. *Ardea* 81: 99–105.

Berg, Å. 1996. Predation on artificial, solitary and aggregated wader nests on farmland. *Oecologia* 107: 343–346.

Berg, Å. 2002. Composition and diversity of bird communities in Swedish farmland-forest mosaic landscapes. *Bird Study* 49: 153–165.

Berg, Å., Lindberg, T. & Kallebrink, K.G. 1992. Hatching success of Lapwings on farmland: differences between habitats and colonies of different sizes. *Journal of Animal Ecology* 61: 469–476.

Berg, Å. & Part, T. 1994. Abundance of breeding farmland birds on arable and set-aside fields at forest edges. *Ecography* 17: 147–152.

Bijlsma, R.G., Meininger, P.L., Rekers, M., de Roder, F.E., Schulting, R. & Vogel, R. 1985. Wader counts in the Tejo estuary near Lisbon and in the Salinas of south Portugal. *Wader Study Group Bulletin* 43: 23–24.

Birdlife International. 2004. *Birds in Europe: population estimates, trends and conservation status.* Cambridge, U.K: Birdlife International. (Birdlife Conservation Series No. 12).

Blomqvist, D. & Johansson, O.C. 1994. Double clutches and uniparental care in Lapwing *Vanellus vanellus,* with a comment on the evolution of double-clutching. *Journal of Avian Biology* 25: 77–79.

Blomqvist, D. & Johansson, O.C. 1995. Trade-offs in nest site selection in coastal populations of Lapwings *Vanellus vanellus. Ibis* 137: 550–558.

Blomqvist, D., Johansson, O.C. & Götmark, F. 1997. Parental quality and egg size affect chick survival in a precocial bird, the Lapwing *Vanellus vanellus. Oecologia* 110: 18–24.

Blondel, J. & Blondel, C. 1964. Remarques sur l'hivernage des limicoles et autres oiseaux aquatiques au Maroc. Janvier 1964. *Alauda* XXXII: 250–279.

Bolam, G. 1912. *The Birds of Northumberland and the Eastern Borders.* H.H. Blair. Alnwick.

Boyle, G.L. 1956. Lapwing chicks 'pattering' and feeding on tadpoles. *British Birds* 49: 502.

Breeding Bird Survey of Japan, 1978. Wild Bird Society of Japan.

Broad, J. 1980. Alternate Husbandry and Permanent Pasture in the Midlands, 1650–1800. *Agricultural History Review* 28(2): 77–89.

Brock, S.E. 1911. Lapwings *Vanellus vanellus* in the pairing season. *Zoologist* series 4, XV: 296–304.

Brouwer, G.A. 1952. Report of documentation on the protection of birds in the Netherlands. *ICBP Bulletin* VI: 176–179.

Brown, A. & Grice, P. 2005. *Birds in England.* T & AD Poyser. London.

Brown, R.H. 1926. Some breeding habits of the Lapwing. *British Birds* 20: 162–168.

Brown, R.H. 1946. Chipping-period of Lapwing's eggs. *British Birds* 39: 124–125.

Browne, M. 1889. *Vertebrate animals of Leicestershire and Rutland.* Privately published.

Brucker, J.W., Gosler, A.G. & Heryet, A.R. 1992. *Birds of Oxfordshire.* Pisces Publications. Newbury.

Buckingham, D.L., Peach, W.J. & Fox, D.S. 2006. Effects of agricultural management on the use of lowland grassland by foraging birds. *Agriculture, Ecosystems and Environment* 112: 21–40.

Buckley, T.E. & Harvie-Brown, J.A. 1891. *A Vertebrate Fauna of the Orkney Islands.* David Douglas. Edinburgh.

Bundy, G. & Warr, E. 1981. A checklist of the birds of the Arabian Gulf States. *Sandgrouse* 1: 4–49.

Burton, J.F. 1995. *Birds and Climate Change.* Christopher Helm. London.

Busche, G. 1994. (The decline of wet meadow birds in Schleswig-Holstein, Germany, from 1950 to 1992). *J.Orn.* 135: 167–177. English summary.

Butler, S.J. & Gillings, S. 2004. Quantifying the effects of habitat structure on prey detectability and accessibility to farmland birds. *Ibis* 146 (Suppl.2) 123–130.

Byrkjedal, I., Grønstøl, G.B., Lislevand, T., Pedersen, K.M., Sandvik, H. & Stalheim, S. 1997. Mating systems and territory in Lapwings *Vanellus vanellus. Ibis* 139: 129–137.

Byrkjedal, I & Thompson, D.B.A. 1998. *Tundra Plovers: The Eurasian, Pacific and American Golden Plovers and Grey Plover.* T & AD Poyser. London.

Calbrade, N., Entwhistle, C.A., Smith, A.J. & Spencer, K.G. 2001. Roof assemblies of lapwings and plovers in Britain. *British Birds* 94: 35–38.

Campbell, B. & Lack, E. (eds). 1985. *A Dictionary of Birds.* T & AD Poyser. Calton.

Carroll, Z.L., Bird, S.B., Emmett, B.A., Reynolds, B. & Sinclair, F.L. (2004). Can tree shelterbelts on agricultural land reduce flood risk? *Soil Use and Management* 20: 357–359.

Catchpole, E.A., Byron, J.T.M., Freeman, S. N & Peach, W.J. 1999. Modelling the survival of British Lapwings *Vanellus vanellus* using ring-recovery data and weather covariates. *Bird Study* 46 (suppl.), S5–13.

Cave, F.O. & MacDonald, J.D. 1955. *The Birds of Sudan.* Oliver & Boyd. Edinburgh.

CAYFORD, J.T. & WATERS, R.J. 1996. Population estimates for waders Charadrii wintering in Great Britain, 1987/88–1991/92. *Biological Conservation* 77: 7–17.
CHALMERS, A.G. & LEECH, P.K. 1986. *Fertiliser use on farm crops in England and Wales 1985.* MAFF. London.
CHAMBERLAIN, D.E. & FULLER, R.J. 2000. Local extinctions and changes in species richness of lowland farmland birds in England and Wales in relation to recent changes in agricultural land-use. *Agriculture, Ecosystems & Environment* 78: 1–17.
CHAMBERLAIN, D.E. & CRICK, H.Q.P. 2003. Temporal and spatial associations in aspects of reproductive performance of Lapwings *Vanellus vanellus* in the United Kingdom, 1962–1999. *Ardea* 91: 183–196.
CHANCE, E.P. 1930. Delayed laying of a Lapwing. *British Birds* 24: 112.
CHEW, H.C. 1953. The post-War land use pattern of the former grasslands of eastern Leicestershire. *Geography* 38: 286–295.
CHISHOLM, M. 1968. *Rural Settlement and Land Use.* Hutchinson University Library. London.
CHRISTY, R.M. 1890. *The Birds of Essex.* Simkin Marshall Hamilton. Chelmsford.
CHURCH, B.M. 1975. *Fertiliser use in farm type regions of England and Wales.* MAFF. London.
CHURCH, B.M. & LEECH, P.K. 1983. *Fertiliser use on farm crops in England and Wales 1982.* MAFF. London.
CLARK, J.M. & EYRE, J.A. 1993. *Birds of Hampshire.* Hampshire Ornithological Society.
CLARKE, T. 2006. *Birds of the Atlantic Islands.* Christopher Helm. London.
COCKER, M. & MABEY, R. 2005. *Birds Britannica.* Chatto & Windus. London.
CONDER, P. 1981. Birds of the Azraq Wetland Reserve, Jordan: January and February 1979. *Sandgrouse* 2: 22–32.
COLLINGE, W.E. 1924–27. *The Food of some British Wild Birds.* Privately published.
COOKE, A.S. 1997. Avermectin use in livestock. Technical information leaflet published by FWAG. Stoneleigh. Warwickshire.
COPPOCK, J.T. 1976. *An Agricultural Atlas of England and Wales (2nd edn.).* Faber & Faber. London.
COTT, H.B. 1953. The exploitation of wild birds for their eggs. *Ibis* 95: 409–449, 643–675 & 96 (1954) 129–149.
COWARD, T.A. & OLDHAM, C. 1904. Notes on the birds of Anglesea. *Zoologist* series 4, 8: 7–29.
CRAMP, S., BOURNE, W.R.P. & SAUNDERS, D. 1974. *The Seabirds of Britain and Ireland.* Collins. London.
CRAMP, S. & SIMMONS, K.E.I. (eds). 1979, 1982. *The Birds of the Western Palaearctic Vols: II & III.* Oxford University Press. Oxford.
CROOKS, S.E. & MOXEY, P.A. 1966. Study of a wintering Lapwing population in North West Middlesex. *London Bird Report* 30(1965): 60–79.
CUCCO, M., LEVI, L., MAFFEI, G. & PULCHER, C. 1996. *Atlante degli uccelli di Piemonte e Valle d'Aosta in inverno 1986–1992.* Museo Regionale di Scienze Naturali. Turin.
DABELSTEEN, T. 1978. An analysis of the song-flight of the Lapwing (*Vanellus vanellus*) with respect to causation, evolution and adaptations to signal function. *Behaviour* 66: 136–178.
DARGIE, T.C. 1993. *The Distribution of Lowland Wet Grassland in England.* English Nature Research Report No.49. English Nature. Peterborough.
DAVIS, T. 1794. *General View of the Agriculture of the County of Wiltshire.* Board of Agriculture. London.
DECEUNINCK, B. & MAHÉO, R. 1998. Limicoles nicheurs de France. *Ornithos* 5: 97–117.
DEL HOYO, J., ELLIOT, A. & SARGATAL, J. (eds). 1996. *The Handbook of the Birds of the World: vol III.* Lynx Edicions. Barcelona.
DEMENTIEV, G.P. & GLADKOV, N.A. (eds). 1969. *Birds of the Soviet Union Vol.III.* Israel Program for Scientific Translation.
DENNIS, M.K. 1996. *A Tetrad Atlas of Breeding Birds of Essex.* Essex Bird-watching and Preservation Society.
DEVEREUX, C.L., MCKEEVER, C.U., BENTON, T.G. & WHITTINGHAM, M.J. 2004. The effect of sward height and drainage on Common Starlings *Sturnus vulgaris* and Northern Lapwings *Vanellus vanellus* foraging in grassland habitats. *Ibis* 146 (Suppl.2): 115–122.

DEVILLERS, P., ROGGEMAN, W., TRICOT, J., DEL MARMOL, P., KERWIJN, C., JACOB, J-P. & ANSELIN, A. 1988. Atlas des Oiseaux nicheurs de Belgique. Institut Royal des Sciences Naturelles de Belgique. Brussels.

DEVOS, K., MEIRE, P. & KUIJKEN, E. 1991. Recent population estimates of meadow breeding waders in Belgium. *Wader Study Group Bulletin* 61, *Supplement:* 14–22.

DIXON, A. 1995. The use of a former opencast coal mine site by breeding Lapwing *Vanellus vanellus. Welsh Birds* 1: 3–10.

DIXON, A. 1996. Lapwing breeding at Waun-y-Pound in 1995. *Welsh Birds* 1(3): 59.

DOBINSON, H.M. & RICHARDS, A.J. 1964. The effects of the severe winter of 1962/63 on birds in Britain. *British Birds* 57: 373–434.

DONALD, P.A. 2004. *The Skylark.* T & AD Poyser. London.

DONOVAN, J. & REES, G. 1994. *Birds of Pembrokeshire.* Dyfed Wildlife Trust. Haverfordwest.

DUBOIS, P.J., MAHÉO, R. & HÖTKER, H. 1991. Waders breeding on wet grasslands in France. *Wader Study Group Bulletin* 61, *Supplement:* 27–31.

DYRCZ, A, WITKOWSKI, J. & OKULEWICZ, J. 1981. Nesting of 'timid' waders in the vicinity of 'bold' ones as an antipredator adaptation. *Ibis* 123: 542–545.

DYRCZ, A. & WITOWSKI, J. 1987. Numbers, distribution and interspecific relations of breeding waders in natural Biebrza fen and adjacent reclaimed marsh. *Wader Study Group Bulletin* 51: 42–44.

EDWARDS, C.A. 1984. Changes in agricultural practice and their impact on soil organisms. In: Jenkins 1984 (ed). *Agriculture and the Environment.* Institute of Terrestrial Ecology. Cambridge.

ELLIOT, R.D. 1982. Dispersion of lapwing nests in relation to predation and anti-predator defence. PhD Thesis. University of Aberdeen.

ELLIOT, R.D. 1985a. The exclusion of avian predators from aggregations of nesting Lapwings (*Vanellus vanellus*). *Animal Behaviour* 33: 308–314.

ELLIOT, R.D. 1985b. The effects of predation risk and group size on the anti-predator responses of nesting Lapwings *Vanellus vanellus. Behaviour* 92: 168–187.

ELLIS, E.A. 1965. *The Broads.* Collins London.

ERIKSSON, M.O.G. & GÖTMARK, F. 1982. Habitat selection: do passerines nest in association with Lapwings *Vanellus vanellus* as defence against predators. *Ornis Scandinavica* 13: 189–192.

ETCHÉCOPAR, R.D. & HÜE, F. 1967. *The Birds of North Africa.* Oliver & Boyd. Edinburgh & London.

ETTRUP, H. & BAK, B. 1985. Nogle traek af danske Vibers *Vanellus vanellus* yngleforhold. *Dansk. Orn. Foren. Tidsskr* 79:43–55.

EUROSTAT 1980. *Land use and production 1955–1979.* Office for Official Publications of the European Communities. Luxemburg.

EUROSTAT 1995. *Statistics in Focus.* Luxemburg. Commission of the European Communities.

EVANS, C., STREET, L., BENSTEAD, P., CADBURY, J., HIRONS, G., SELF, M. & WALLACE, H. 1995. Water and sward management for Nature Conservation: a case study of the RSPB's West Sedgemoor reserve. *RSPB Conservation Review No.9* 60–72.

EVANS, W. 1908. Letter. *Annals of Scottish Natural History.* 67: 255.

EVERETT, M.J. 1987. The Elmley Experiment. *RSPB Conservation Review No.1* 31–33.

FAGAN, B. 2000. *The Little Ice Age: how climate made history 1300–1850.* Basic Books. New York.

FIELD, J. 1993. *A History of English Field-names.* Longman. Harlow.

FINLAYSON, C. 1992. *The Birds of the Straits of Gibraltar.* T & AD Poyser. London.

FLETCHER, K., WARREN, P, & BAINES, D. 2005. Impacts of nest visits by human observers on hatching success in Lapwings *Vanellus vanellus*: a field experiment. *Bird Study* 52: 221–223.

FORNASARI, L., BOTTONI, L. & RENATO, M. (eds). 1992. *Atlante degli uccelli svernanti in Lombardia.* University of Milan. Milan.

FOX, A.D. 2004. Has Danish agriculture maintained farmland bird populations? *Journal of Applied Ecology* 41: 427–439.

FRIKKE, J. 1991. Breeding waders and wet grassland habitats in Denmark. *Wader Study Group Bulletin* 61, *Supplement*: 42–49.

FULLER, R.J. 1983. The relative growth of bill, tarsus and wing of Lapwing chicks in the Outer Hebrides. *Ringing and Migration* 4: 139–144.

FULLER, R.J. 1986. *Lapwing*. In: Lack, P. 1986. *The Atlas of Wintering birds in Britain and Ireland.* T & AD Poyser. Calton.

FULLER, R.J. 1988. Wintering Golden Plovers in central Buckinghamshire: annual variation in numbers and distribution. *Buckinghamshire Bird Report 1988*: 4–8.

FULLER, R.J. 1993. *Lapwing*. In: Lack. P & Ferguson D. 1993. *The Birds of Buckinghamshire.* Buckinghamshire Bird Club.

FULLER, R.J. 1994. Changes in breeding populations of Curlews and Lapwings in central Buckinghamshire, 1981–1990. *Buckinghamshire Bird Report 1994* 11–16.

FULLER, R.J. & YOUNGMAN, R.E. 1979. The utilization of farmland by Golden Plovers wintering in southern England. *Bird Study* 26: 37–46.

FULLER, R.J. & LLOYD, D. 1981. The distribution and habitats of wintering Golden Plovers in Britain, 1977–1978. *Bird Study* 28: 169–185.

FULLER, R.J., REED, T.M., BUXTON, N.E., WEBB, A., WILLIAMS, T.D. & PIENKOWSKI, M.W. 1986. Populations of breeding waders Charadrii and their habitats on the crofting lands of the Outer Hebrides, Scotland. *Biological Conservation* 37: 333–361.

FULLER, R.J. & GOUGH, S.J. 1996. *Relationships between Grazing and Birds with particular reference to Sheep in the British Uplands.* BTO Research Report No.164. Thetford.

FULLER, R.J., WARD, E., HIRD, D. & BROWN, A.F. 2002. Declines of ground-nesting birds in two areas of upland farmland in the south Pennines of England. *Bird Study* 49: 146–152.

FURBANK, P.N., OWENS, W.R. & COULSON, A.J. 1991. *Daniel Defoe: A Tour Through the Whole Island of Great Britain.* Yale University Press. London.

GALBRAITH, H. 1987. Threats to breeding waders: the impact of changing agricultural land-use on the breeding ecology of Lapwings. *Wader Study Group Bulletin* 49, *Supplement*: 102–104.

GALBRAITH, H. 1988a. Effects of agriculture on the breeding ecology of Lapwings *Vanellus vanellus. Journal of Applied Ecology* 25: 487–503.

GALBRAITH, H. 1988b. Effects of egg size and composition on the size, quality and survival of lapwing *Vanellus vanellus* chicks. *Journal of the Zoological Society of London* 214: 383–398.

GALBRAITH, H. 1988c. The effects of territorial behaviour on Lapwing populations. *Ornis Scandinavica* 19: 134–138.

GALBRAITH, H. 1988d. Adaptation and constraint in the growth pattern of lapwing *Vanellus vanellus* chicks. *Journal of the Zoological Society of London* 215: 537–548.

GALBRAITH, H. 1989a. The diet of Lapwing *Vanellus vanellus* chicks on Scottish farmland. *Ibis* 131: 80–84.

GALBRAITH, H. 1989b. Arrival and habitat use by Lapwings *Vanellus vanellus* in the early breeding season. *Ibis* 131: 377–388.

GALBRAITH, H. & FURNESS, R.W. 1983. Breeding waders on agricultural land. *Scottish Birds* 12: 148–153.

GALBRAITH H., FURNESS, R.W. & FULLER, R.J. 1984. Habitats and distribution of waders breeding on Scottish agricultural land. *Scottish Birds* 13: 98–107.

GALLAGHER, M. & WOODCOCK, M.W. 1980. *Birds of Oman*. Quartet Books. London.

GIBBONS, D.W., REID, J.B. & CHAPMAN, R.A. 1993. *The New Atlas of Breeding Birds in Britain and Ireland: 1988–1991.* T & AD Poyser. London.

GILL, J.A. 1996. Habitat choice in pink-footed geese: quantifying the constraints determining winter site use. *Journal of Applied Ecology* 33: 884–892.

GILLINGS, S. 1999. *Lapwing*: in Taylor, M., Seago, M.J., Allard, P. & Dorling, D. 1999. *The Birds of Norfolk.* Pica Press. Mountfield.

GILLINGS, S. 2003. *Diurnal and nocturnal ecology of Golden Plovers Pluvialis apricaria and Lapwings Vanellus vanellus wintering on arable farmland.* PhD Thesis to the University of East Anglia.

GILLINGS, S. & BEAVEN, P. 2004. Wintering farmland birds: results from mass-participation surveys. *British Birds* 97: 118–129.

GILLINGS, S., FULLER, R.J. & SUTHERLAND, W.J. 2005. Diurnal studies do not predict nocturnal habitat choice and site selection of European Golden Plovers (*Pluvialis apricaria*) and Northern Lapwing (*Vanellus vanellus*). *The Auk* 122: 1249–1260.

Gillings, S., Austin, G.E., Fuller, R.J. & Sutherland, W.J. 2006. Distribution shifts in wintering Golden Plovers *Pluvialis apricaria* and Lapwings *Vanellus vanellus* in Britain. *Bird Study* 53: 274–284.

Gillings, S., Fuller, R.J. & Sutherland, W.J. in press. Winter field use and habitat selection by Eurasian Golden Plovers *Pluvialis apricaria* and Northern Lapwings *Vanellus vanellus* on arable farmland. *Ibis.*

Gjershaug, J.O., Thingstad, P.G., Eldøy, S. & Byrkjeland, S. 1994. *Norsk Fugleatlas.* Norsk Ornitologisk Forening.

Gladstone, H.S. 1910. *The Birds of Dumfriesshire.* Witherby. London.

Glutz von Blotzheim, U.N., Bauer, K.M. & Bezzel, E. 1975. *Handbuch der Vogel Mitteleuropas Vol. 6.* Akademische Verlagsgesellschaft AULA-Verlag. Weisbaden.

Godfrey, W.E. 1966. *The Birds of Canada.* National Museum of Canada. Bulletin 203.

Goodman, S.M. & Meininger, P.L. (eds). 1989. *The Birds of Egypt.* University Press. Oxford.

Gorman, G. 1996. *The Birds of Hungary.* Helm. London.

Graul, W.D. 1973. Possible functions of head and breast markings in Charadriinae. *Wilson Bulletin* 85: 60–70

Green, R.E. & Robins, M. 1993. The decline of the ornithological importance of the Somerset Levels, England, and changes in the management of water levels. *Biological Conservation* 66: 95–106.

Gregory, R.D. 1987. Comparative winter feeding ecology of Lapwings *Vanellus vanellus* and Golden Plovers *Pluvialis apricaria* on cereals and grasslands in the Lower Derwent Valley, North Yorkshire. *Bird Study* 34: 244–250.

Gregory, R.D. & Marchant, J.H. 1996. Population trends of Jays, Magpies, Jackdaws and Carrion Crows in the United Kingdom. *Bird Study* 43: 28–37.

Grigg, D. 1989. *English Agriculture: An Historical Perspective.* Basil Blackwell. Oxford.

Grimmett, R., Inskipp, C. & Inskipp, T. 1998. *The Birds of the Indian Subcontinent.* Helm. London.

Gromadzka, J., Stawarczyk, T. & Tomialojc, L. 1985. Breeding waders in Poland. *Wader Study Group Bulletin* 43: 29–33.

Grønstøl, G.B. 1996. Aerobatic components in the song-flight display of male Lapwings *Vanellus vanellus* as cues in female choice. *Ardea* 84: 45–55.

Gudmundsson, F. 1951. The effect of the recent climatic changes on bird-life in Iceland. *Proc. Xth I.O.C* 502–514.

Haartman, L. von. 1973. Changes in the breeding bird fauna of North Europe: In Farner, D.S. (ed). *Breeding Biology of Birds.* National Academy of Sciences, Washington.

Hafsmo, J.E., Brykejedal, I., Grønstøl, G.B. & Lislevand, T. 2001. Simultaneous tetragyny in Northern Lapwing *Vanellus vanellus. Bird Study* 48: 124–125.

Haftorn, S. 1958. (Population changes, especially geographical changes, in the Norwegian avifauna during the last 100 years). *Sterna* 3: 105–137. English summary.

Hagemeijer, W.J.M. & Blair, M.J. (eds). 1997. *The EBCC Atlas of European Breeding Birds.* T & AD Poyser. London.

Hakala, A.V.K. 1971. A quantitative study of the bird fauna of some open peatlands in Finland. *Ornis Fennica* 48: 1–11.

Haldane, J.B.S. 1955. The calculation of mortality rates from ringing data. In: Portmann, A. & Sutter, E. (eds). *Proceedings of the Eleventh International Ornithological Congress.* Birkhauser Verlag. Basel and Stuttgart.

Hale, W.G. 1980. *Waders.* Collins N/N. London.

Hampson, P., Beaumont, D. & Peacock, M. 1996. Integrating farming and wildlife conservation at the RSPB Loch Gruinart nature reserve. *RSPB Conservation Review* 10: 59–68.

Harris, G. 1964. Climatic changes since 1860 affecting European Birds. *Weather* 19: 70–79.

Harrison, M.J., Mead, W.R. & Pannett, D.J. 1965. A Midland ridge-and-furrow map. *Geographical Journal.* 131, 3: 366–369.

Hart, J.D., Milsom, T.P., Baxter, A., Kelly, P.F. & Parkin, W.K. 2002. The impact of livestock on Lapwing *Vanellus vanellus* breeding densities and performance on coastal grazing marsh. *Bird Study* 49: 67–78.

HARTHAN, A.J. 1946. *The Birds of Worcestershire.* Littlebury & Co. Worcester.

HARVIE-BROWN, J.A. 1906. *A Fauna of the Tay Basin and Strathmore.* David Douglas. Edinburgh.

HARVIE-BROWN, J.A. & BUCKLEY, T.E. 1895. *A Fauna of the Moray Basin.* David Douglas. Edinburgh.

HARVIE-BROWN, J.A. & MACPHERSON, H.A. 1904. *A Fauna of the North-west Highlands and Skye.* David Douglas. Edinburgh.

HASLAM, S.M. 1991. *The Historic River: rivers and culture down the ages.* Cobden of Cambridge.

HAVERSCHMIDT, F. 1943. De Goudpluvierenvangst in Nederland. *Ardea* 32: 35–74.

HAVILAND, M.D. 1915. Notes on the courtship of the Lapwing. *Zoologist* ser.4. XIX: 217–225.

HAYMAN, P., MARCHANT, J. & PRATER A. 1986. *Shorebirds. An identification guide to the waders of the world.* Croom Helm. London.

HEATHCOTE, A., SALMON, H.M. & GRIFFIN, D. 1967. *The Birds of Glamorgan.* Cardiff Naturalists' Society. Cardiff.

HEGYI, Z. & SASVÁRI, L. 1998. Components of fitness in Lapwings *Vanellus vanellus* and Black-tailed Godwits *Limosa limosa* during the breeding season: do female body mass and egg size matter? *Ardea* 86: 43–50.

HEIM, P.J. 1974. Eiablage, Gelegegrosse und Brutdauer beim Kiebitz *Vanellus vanellus. Der Ornithologische Beobachter* 71: 283–288.

HEIM, P.J. 1978. Populationskologische Daten der Nuoler Kiebitzkolonie *Vanellus vanellus,* 1948–1977. *Der Ornithologische Beobachter* 75: 85–94.

HENDERSON, I.G., WILSON, A.M., STEELE, D. & VICKERY, J.A. 2002. Population estimates, trends and habitat associations of breeding Lapwing *Vanellus vanellus,* Curlew *Numenius arquata* and Snipe *Gallinago gallinago* in Northern Ireland in 1999. *Bird Study* 49: 17–25.

HENDERSON, I.G., BODELY, T. & CLARK, N. 2004a. Arable options for breeding Lapwings. *BTO News* 250: 14.

HENDERSON, I.G., FULLER, R.J., CONWAY, G.J. & GOUGH, S.J. 2004b. Evidence for declines in populations of grassland-associated birds in marginal upland areas of Britain. *Bird Study* 51: 12–19.

HICKLING, R.A.O. 1967. The inland wintering of gulls in England, 1963. *Bird Study* 14: 104–113.

HILDÉN, O. 1965. Habitat selection in birds: A review. *Ann. Zool. Fenn.* 2: 53–70.

HÖGSTEDT, G. 1974. The length of the pre-laying period in the lapwing *Vanellus vanellus* in relation to its food resources. *Ornis Scandinavica* 5: 1–4.

HOLLAND, P., SPENCE, I. & SUTTON, I. 1984. *Breeding Birds in Greater Manchester.* Manchester Ornithological Society.

HOLLOWAY, S. 1996. *The Historical Atlas of Breeding Birds in Britain and Ireland 1875–1900.* T & AD Poyser. London.

HOSKINS, W.G. 1955. *The Making of the English Landscape.* Hodder & Stoughton. London.

HÖTKER, H. 1991. Waders breeding on wet grasslands in the countries of the European Community—a brief summary of current knowledge on population sizes and population trends. *Wader Study Group Bulletin* 61, *Supplement*: 50–55.

HOWARD, E. 1920. *Territory in Bird Life.* Collins. London.

HROMADKOVA, V. 1987. The distribution of breeding waders in Czechoslovakia in 1973–1977. *Wader Study Group Bulletin* 50: 24–27.

HÜE, F. & ETCHÉCOPAR, R.D. 1970. *Les Oiseaux du Proche et du Moyen Orient.* Paris.

HURFORD, C. & LANSDOWN, P. 1995. *Birds of Glamorgan.* Published by the authors. Cardiff.

HUSTINGS, F. 1997. The decline of the Corn Bunting *Miliaria calandra* in the Netherlands. In: Donald, P.F. & Aebischer, N.J. (eds). *The ecology and conservation of Corn Buntings Miliaria calandra.* UK Conservation No: 13. JNCC. Peterborough.

HUTCHINSON, C.D. 1989. *Birds in Ireland.* T & AD Poyser. Calton.

IAPICHINO, C. & MASSA, B. 1989. *The Birds of Sicily.* BOU Check List No.11. British Ornithologists Union. Tring.

IMBODEN, C. 1970. Zur okologie einer randzonen-population des Kiebitz in der Schweiz. *Der Ornithologische Beobachter* 67: 41–58.

IMBODEN, C. 1971. Bestand, verbreitung und biotop des Kiebitz in der Schweiz. *Der Ornithologische Beobachter* 68: 37–53.

IMBODEN, C. 1974. Zug, Fremdansiedlung und Brutperiode des Kiebitz *Vanellus vanellus* in Europa. *Der Ornithologische Beobachter* 71: 5–134.

INSKIPP, C. & INSKIPP, T. 1985. *A Guide to the Birds of Nepal.* Christopher Helm. London.

ISRAEL, J. 1995. *The Dutch Republic: its Rise, Greatness and Fall, 1477–1806.* University Press. Oxford.

IVERSEN, F.M. 1986. (The impact of disturbance on the Lapwing's *Vanellus vanellus* incubation) In Danish with English summary. *Dansk Orn. Foren. Tidsskr.* 80: 97–102.

JACKSON, D.B. & GREEN, R.E. 2000. The importance of the introduced Hedgehog (*Erinaceus europaeus*) as a predator of the eggs of waders (Charadrii) on machair in South Uist, Scotland. *Biological Conservation* 93: 333–348.

JACKSON, D.B., FULLER, R.J. & CAMPBELL, S.T. 2004. Long-term population changes among breeding shorebirds in the Outer Hebrides, Scotland, in relation to introduced Hedgehogs (*Erinaceus europaeus*). *Biological Conservation* 117: 151–166.

JACKSON, R. & JACKSON, J. 1975. A study of breeding Lapwings in the New Forest, Hampshire. *Ringing and Migration* 1: 18–27.

JACKSON, R. & JACKSON, J. 1980. A study of Lapwing breeding population changes in the New Forest, Hampshire. *Bird Study* 27: 27–34.

JENNINGS, M.C. 1981. *The Birds of Saudi Arabia: a checklist.* Privately published.

JESSE, R.H.B. 1960. *A Survey of the Agriculture of Sussex.* Royal Agricultural Society of England. London.

JOHANSSON, O.C. & BLOMQVIST, D. 1996. Habitat selection and diet of Lapwing *Vanellus vanellus* chicks on coastal farmland in SW Sweden. *Journal of Applied Ecology* 33: 1030–1040.

JOURDAIN, F.C.R. & WITHERBY, H.F. 1918. The effects of the winter of 1916/17 on our resident birds. *British Birds* 11: 266–271 & 12: 26–35.

JUNE CENSUS STATISTICS. Published annually by Ministry of Agriculture, now Department of the Environment, Fisheries and Rural Affairs.

KALELA, O. 1949. Changes in geographic ranges in the avifauna of northern and central Europe in relation to recent changes in climate. *Bird Banding* 20: 77–103.

KALLANDER, H. 1977. Piracy by Black-headed Gulls on Lapwings. *Bird Study* 24: 186–194.

KELSALL, J.E. & MUNN, P.W. 1905. *The Birds of Hampshire and the Isle of Wight.* Witherby. London.

KENNEDY, P.G., RUTTLEDGE, R.F. & SCROOPE, C.F. 1954. *Birds of Ireland.* Oliver & Boyd. Edinburgh and London.

KERSTEN, M. & SMIT, C.J. 1984. The Atlantic coast of Morocco. In: Evans, P.R. Goss-Custard, J.D. & Hale, W.G. 1984. *Coastal waders and wildfowl in winter.* University Press. Cambridge.

KIRBY, J.S. 1997. Influence of environmental factors on the numbers and activity of wintering Lapwings and Golden Plovers. *Bird Study* 44: 97–110.

KIRBY, J.S. & LACK, P.C. 1993. Spatial dynamics of wintering Lapwings and Golden Plovers in Britain and Ireland, 1981/82–1983/84. *Bird Study* 40: 38–50.

KIRBY, J.S. & FULLER, R.J. 1994. Winter distribution and habitat use by lapwings *Vanellus vanellus* in the Vale of Aylesbury. In: Tucker, G.M., Davies, S.M. & Fuller, R.J. (eds). *The ecology and conservation of lapwings Vanellus vanellus.* UK Conservation No. 9. JNCC. Peterborough.

KIRKMAN, F.B. & HUTCHINSON, H.G. 1924. *British Sporting Birds.* T.C. & E.E. Jack. London and Edinburgh.

KLINNER, B. 1991. Breeding waders on wet grasslands (inland sites) in west Germany: recent data. *Wader Study Group Bulletin* 61, *Supplement*: 22–25.

KLOMP, H. 1951. Over de achteruitgang van de Kievit *Vanellus vanellus* (L) in Nederland en gegevens over het legmechanisme en het eiproductie-vermogen. *Ardea* 39: 143–182. English summary.

KLOMP, H. 1954. De terreinkuis van de Kievit *Vanellus vanellus* (L). *Ardea* 42: 1–139.

KLOMP, H. & SPEEK, B.J. 1971. Survival of young Lapwings in relation to time of hatching. *Bird Study* 18: 229–231.

KNIGHT, J. 1988. Wintering Golden Plovers and Lapwings in Buckinghamshire in December 1986. *Buckinghamshire Bird Report.*

KNIGHT, R., SCROGGS, R.W.H. & WILLIAMS, H.H. 1980. Annual Breeding Season Survey—1980. The Lapwing *Vanellus vanellus. Banbury Ornithological Society Report.*

KNUBLEY, E.P. 1892. Is legislative protection required for wild birds' eggs? *Annals of Scottish Natural History* 4: 213–218.

KOOIKER, G. 1984. Brutokologische untersuchungen an einer population des Kiebitzes. *Die Vogelwelt* 105: 121–137.

KOOIKER, G. 1987. Gelegegrosse, Schlupfrate, Schlupferfolg und Bruterfolg beim Kiebitz *Vanellus vanellus*. *Journal fur Ornithologie* 128: 101–107.

KOOIKER, G. 1990. Bestandsentwicklung und Bruterfolg einer Kiebitzpopulation *Vanellus vanellus* im Agrarraum bei Osnabruck. *Die Vogelwelt* 111: 202–215.

KOSKIMIES, P. 1989. Distribution and numbers of Finnish breeding birds. Appendix to *Suomen Lintuatlas*. SLY:n Lintutieto Oy. Helsinki.

KOSKIMIES, P. 1992. Population sizes and recent trends of breeding birds in the Nordic countries. *Bird Census News* 41–79.

KRUK, M., NOORDERVLIET, M.A.W. & TER KEURS, W.J. 1996. Hatching dates of waders and mowing dates in intensively exploited grassland areas in different years. *Biological Conservation* 77: 213–218.

KVAERNE, M. 1973. (Is the Lapwing a 'watch-dog' species for other wader species.) *Sterna* 12: 85–90. English summary.

LACK, D. 1947. The significance of clutch size. *Ibis* 89: 302–352.

LACK, P. 1986. *The Atlas of Wintering birds in Britain and Ireland*. T & AD Poyser. Calton.

LACK, P. & FERGUSON, D. 1993. *The Birds of Buckinghamshire*. Buckinghamshire Bird Club. High Wycombe.

LAIDLAW, T.G. 1908. Food of the Black-headed Gull. *Annals of Scottish Natural History* 67: 139–141.

LARSEN, L.G. 1987. The status and protection of freshwater meadows and their bird-life in Denmark. *Zoologica* 14: 169–175. *Proceedings of the Fifth Nordic Ornithological Congress*, 1985.

LARSSON, T. 1969. Land use and bird faunas on shores in south Sweden. *Oikos* 20: 136–155.

LARSSON, T. 1976. Composition and density of the bird fauna in Swedish shore meadows. *Ornis Scandinavica* 7: 1–12.

LAVEN, B. 1941. Beobachtungen uber Balz und Brut biem Kiebitz (*Vanellus vanellus*). *Journal fur Ornithologie Erganzungsband* 3: 1–64.

LEA, D. & BOURNE, W.R.P. 1975. The birds of Orkney. *British Birds* 68: 261–283.

LEECH, D. & CRICK, H.Q.P. 2005. Nest Record Scheme breeding trends—latest results. *BTO News* 261: 18–19.

LEFRANC, N. 1997. Shrikes and the farmed landscape in France. In: Pain, D.J. & Pienkowski, M.W. (eds). 1997. *Farming and Birds in Europe*. Academic Press. London.

LEFRANC, N. & WORFOLK, T. 1997. *Shrikes: a Guide to the Shrikes of the World*. Pica Press. Mountfield.

LEIBAK, E., LILLELEHT, V. & VEROMANN, H. 1994. *Birds of Estonia; status, distribution and numbers*. Estonian Academy of Sciences. Tallinn.

LEROUX, A.B.A. 1991. Drainage of wet meadows and birds in West Marshes of France. In: Curtis, D.J., Bignal, E.M. & Curtis, M.A. (eds). *Birds and Pastoral Agriculture in Europe*. Scottish Chough Study Group. JNCC Peterborough.

LIKER, A. 1992a. A bibic (*Vanellus vanellus*) szaporodasbiologiai vizsgalata szikespusztai elohelyen. *Ornis Hungarica* 2: 61–66.

LIKER, A. 1992b. Breeding biology of Lapwings *Vanellus vanellus* in Southern Hungary. *Wader Study Group Bulletin* 66: 38.

LIKER, A. & SZÉKELY, T. 1997. Aggression between female Lapwings *Vanellus vanellus*. *Animal Behaviour* 54: 797–802.

LIKER, A. & SZÉKELY, T. 1999a. Mate choice in Lapwing *Vanellus vanellus*: the significance of song flight behaviour. *Wader Study Group Bulletin* 88: 10.

LIKER, A. & SZÉKELY, T. 1999b. Parental behaviour in the Lapwing *Vanellus vanellus*. *Ibis* 141: 608–614.

LILFORD, LORD. 1895. *Notes on the Birds of Northamptonshire and Neighbourhood*. R.H. Porter. London.

LIND, A.H. 1957. Territorial opforsel hos Vibe (*Vanellus vanellus* L) om efteraret. *Dansk Ornitologisk Forenings Tidsskrift* 51: 22–29.

LISTER, M.D. 1939. An account of the Lapwing population of a Surrey farm. *British Birds* 32: 260–271.

LISTER, M.D. 1964. The Lapwing habitat enquiry, 1960–61. *Bird Study* 11: 128–147.

LLOYD, H.G. 1980. *The Red Fox*. Batsford. London.

LLOYD, L.R.W. 1912. Moving Lapwings' nests. *British Birds* 6: 162–163.

LOCK, L. 1998. *Review of farmland birds and lowland pastoral systems in SW England.* RSPB unpublished report.

LOVEGROVE, R, WILLIAMS, G. & WILLIAMS, I. 1994. *Birds in Wales.* T & AD Poyser. London.

LOVEGROVE, R., SHRUBB, M. & WILLIAMS, I. 1995. *Silent Fields: the current status of farmland birds in Wales.* RSPB.

LUBBOCK, REV. R. 1845. *Observations on the fauna of Norfolk.* Longmans. London.

LUBBOCK, REV. R. 1879. *Observations on the Fauna of Norfolk.* 2nd edition edited by T. Southwell. Jarrold & Sons. Norwich.

LYNES, H. 1910. Manoeuvres of Lapwing in defence of young. *British Birds* 4: 157.

MACGILLIVRAY, W. 1837–52. *History of British Birds.* Scott, Webster and Geary. London.

MACKINNON, J. & PHILLIPS, K. 2000. *A Field Guide to the Birds of China*. University Press. Oxford.

MACPHERSON, H.A. 1897. *History of Fowling: An account of devices for capturing wild birds.* David Douglas. Edinburgh.

MADDEN, B., COONEY, T., O'DONOGHUE, A., NORRISS, D.W. & MERNE, O.J. 1998. Breeding waders of machair systems in Ireland in 1996. *Irish Birds* 6: 177–190.

MAFF. 1968. *A century of agricultural statistics: Great Britain 1866–1966.* HMSO. London.

MARCHANT, J.H., HUDSON, R., CARTER, S.P. & WHITTINGDON, P. 1990. *Population trends in British Breeding Birds.* British Trust for Ornithology. Tring.

MARTÍ, R. & DEL MORAL, J.C. (eds.). 2003. *Atlas de las Aves Reproductoras de España.* Dirección General de Conservación de la Naturaleza–Sociedad Española de Ornitología: Madrid.

MASON, C.F. & MACDONALD, S.M. 1999a. Estuarine feeding by Lapwings *Vanellus vanellus* and Golden Plovers *Pluvialis apricaria. Wildfowl* 50: 205–207.

MASON, C.F. & MACDONALD, S.M. 1999b. Habitat use by Lapwings and Golden Plovers in a largely arable landscape. *Bird Study* 46: 89–99.

MATTER, H. 1982. Einfluß intensiver Feldbewirtschaftung auf den Bruterfolg des Kiebitzes *Vanellus vanellus* in Mitteleuropa. *Der Ornithologische Beobachter* 79: 1–24. English summary.

MAYO, A.L.W. 1974. Wader nesting associations. *British Birds* 67: 82.

MCCRACKEN, D.I. 1993. The potential for avermectins to affect wildlife. *Veterinary Parasitology* 48: 273–280.

MCCRACKEN, D.I. 1995. Short note on potential environmental problems of cattle and sheep anthelmintics. *Extensive Farming Systems Briefing Paper 7.* JNCC. Paisley.

MCCRACKEN, D. & BIGNAL, I. 1991. Chemical alternatives to treatment of cattle with ivermectin. Joint Nature Conservation Committee 1991.

MCCRACKEN, D.I. & FOSTER G.N. 1993. The effect of ivermectin on the invertebrate fauna associated with cow dung. *Environmental Toxicology and Chemistry* 12: 73–84.

MEAD, C.J., FLEGG, J.J. & COX, C.J. 1968. A factor inhibiting subspecific differentiation in the Lapwing. *Bird Study* 15: 105–106.

MEEUS, J.H.A., WIJERMANS, M.P. & VROOM, M.J. 1990. Agricultural landscapes in Europe and their transformation. *Landscape and Urban Planning* 18: 289–352.

MELLERSH, W.L. 1902. *A Treatise on the Birds of Gloucestershire.* Bellows. Gloucester.

MELTOFTE, H. 1980. Wader counts in the Danish part of the Wadden Sea 1974–78. *Fredningsstyrelsen* Copenhagen.

MELTOFTE, H., BLEW, J., FRIKKE, J., ROSNER, H-U & SMIT, C.J. 1994. Numbers and distribution of water birds in the Wadden Sea. *Wader Study Group Bulletin* 74 special issue.

MERIKALLIO, E. 1958. *Finnish Birds: their distribution and numbers.* Societas pro Fauna et Flora Fennica. Helsinki.

MEYER DE SCHAUENSEE, R. 1984. *The Birds of China.* University Press. Oxford.

MILSOM, T.P. 2005. Decline of Northern Lapwing *Vanellus vanellus* breeding in arable farmland in relation to loss of spring tillage. *Bird Study* 52: 297–306.

MILSOM, T.P., HOLDITCH, R.S. & ROCHARD, J.B.A. 1985. Diurnal use of an airfield and adjacent agricultural habitats by Lapwings *Vanellus vanellus*. *Journal of Applied Ecology* 22: 313–326.

MILSOM, T.P., ROCHARD, J.B.A. & POOLE, S.J. 1990. Activity patterns of Lapwings *Vanellus vanellus* in relation to the lunar cycle. *Ornis Scandinavica* 21: 147–156.

MILSOM, T.P., ENNIS, D.C., HASKELL, D.J., LANGTON, S.D. & MCKAY, H.V. 1998. Design of grassland feeding areas for waders during winter: the relative importance of sward, landscape factors and human disturbance. *Biological Conservation* 84: 119–129.

MILSOM, T.P., LANGTON, S.D., PARKIN, W.K., PEEL, S., BISHOP, J.D., HART, J.D. & MOORE, N.P. 2000. Habitat models of bird species' distribution: an aid to the management of coastal grazing marshes. *Journal of Applied Ecology* 37: 706–727.

MILSOM, T.P., HART, J.D., PARKIN, W.K. & PEEL, S. 2002. Management of coastal grazing marshes for breeding waders: the importance of surface topography and wetness. *Biological Conservation* 103: 199–207.

MITCHELL, F.S. 1885. *The Birds of Lancashire*. Van Voorst. London.

MØLLER, A.P. 1983. Changes in Danish farmland habitats and their populations of breeding birds. *Holarctic Ecology* 6: 95–100.

MØLLER, H.S. 1975. Danish salt-marsh communities of breeding birds in relation to different types of management. *Ornis Scandinavica* 6: 125–133.

MOORE, H.J. 1956. *Field observations on the Birds of Iraq*. Iraq Natural History Museum. Baghdad.

MOSS, B. 2001. *The Broads: the people's wetland*. HarperCollins. London.

MURDOCH, D.A., VOS, R., ABDALLAH, A., ABDALLAH, M., ANDREWS, I., AL-ASSAD, A., VAN BEUSEKOM, R., HOFLAND, R., ROTH, T., SAVEYN, B., SERRA, G. & WELLS, C. 2005. A Winter Survey of Syrian Wetlands: final report of the Syrian Wetlands expedition January–February 2004. Privately published.

MURTON, R.K. & WESTWOOD, N.J. 1974. Some effects of agricultural change on the English avifauna. *British Birds* 67: 41–69.

MUIRHEAD, G. 1889. *The Birds of Berwickshire*. David Douglas. Edinburgh.

NAIRN, R.G.W., HERBERT, I.J. & HEERY, S. 1988. Breeding waders and other wet grassland birds of the River Shannon Callows, Ireland. *Irish Birds* 3: 521–537.

NANKINOV, D. 1989. The status of waders in Bulgaria. *Wader Study Group Bulletin* 56: 16–25.

NELSON, T.H. 1907. *The Birds of Yorkshire*. Brown & Sons. London.

NETHERSOLE-THOMPSON, D. 1940. Threat and injury feigning display of Lapwing. *British Birds* 34: 138–139.

NETHERSOLE-THOMPSON, D. & NETHERSOLE-THOMPSON, M. 1986. *Waders, their breeding haunts and watchers*. T & AD Poyser. Calton.

NEWBURY, P.A.R. 1980. *A Geography of Agriculture*. MacDonald & Evans. Plymouth.

NEWTON, A. 1895. *A Dictionary of Birds*. A. & C. Black. London.

NEWTON, I. 1986. *The Sparrowhawk*. T & AD Poyser. Calton.

NEWTON, I. 1998. *Population limitation in Birds*. Academic Press. London.

NEWTON, I. 2003. *The Speciation & Biogeography of Birds*. Academic Press. London.

NICHOLSON, E.M. 1938/39. Report on the Lapwing habitat enquiry. *Brit. Birds* 32: 170–191, 207–229, 255–259.

NICHOLSON, H.H. 1943. Modern field drainage. *Journal of the Royal Agricultural Society of England* 104: 118–135.

OAKES, C. 1953. *The Birds of Lancashire*. Oliver & Boyd. Edinburgh.

O'BRIEN, M. 1996. The numbers of breeding waders in Lowland Scotland. *Scottish Birds* 18: 231–241.

O'BRIEN, M. 2002. The relationship between field occupancy rates by breeding Lapwing and habitat management on upland farmland in Northern Britain. *Aspects of Applied Biology* 67: 85–92.

O'BRIEN, M. & SMITH, K.W. 1992. Changes in the status of waders breeding on wet lowland grasslands in England and Wales between 1982 and 1989. *Bird Study* 39: 165–176.

O'BRIEN, M., THARME, A. & JACKSON, D. 2002. Changes in breeding wader numbers on Scottish farmed land during the 1990s. *Scottish Birds* 23: 10–21.

O'Brien, M. & White, C.S. 2003. Estimating the breeding wader populations of Scottish uplands and salt marshes. *Scottish Birds* 24: 17–28.

O'Connor, R.J. & Shrubb, M. 1986. *Farming & Birds*. Cambridge University Press. Cambridge.

OST Bird Report 1970–73. Ornithological Society of Turkey. Sandy.

Ottvall, R. 2005. (Nest survival among waders breeding on coastal meadows: the relative importance of predation and trampling damage by livestock). In Swedish with English summary. *Ornis Svecica* 15: 89–96.

Parish, D.M.B., Thompson, P.S. & Coulson, J.C. 1997a. Mating systems in the Lapwing *Vanellus vanellus*. *Ibis* 139: 138–143.

Parish, D.M.B., Thompson, P.S. & Coulson, J.C. 1997b. Attempted double-brooding in the Lapwing *Vanellus vanellus*. *Bird Study* 44: 111–113.

Parr, D. (ed). 1972. *Birds in Surrey, 1900–1970*. Batsford. London.

Parr, R. 1993. Nest predation and numbers of Golden Plovers *Pluvialis apricaria* and other moorland waders. *Bird Study* 40: 223–231.

Partridge, J.K. & Smith, K.W. 1992. Breeding wader populations in Northern Ireland, 1985–87. *Irish Birds* 4: 497–518.

Paton, E.R. & Pike, O.G. 1929. *The Birds of Ayrshire*. Witherby. London.

Payn, W.H. 1962. *The Birds of Suffolk*. Barrie & Rockliff. London.

Payne-Gallwey, R. 1882. *The Fowler in Ireland*. Van Voorst. London.

Peach, W.J., Thompson, P.S. & Coulson, J.C. 1994. Annual and long term variation in the survival rates of British Lapwings *Vanellus vanellus*. *Journal of Animal Ecology*. 63: 60–70.

Peers, M. 2003. The decline of the Lapwing in Breconshire. *Breconshire Birds* 2003: 31–35.

Peers, M. & Shrubb, M. 1990. *Birds of Breconshire*. Brecknock Wildlife Trust. Brecon.

Pell, A. 1887. The making of the land in England: a retrospect. *Journal of the Royal Agricultural Society of England* series 2, 23: 355–374.

Pennington, M.G. 1992. Predation of birds' eggs and chicks by herbivorous animals. *Scottish Birds* 16: 285.

Piersma, T., Rogers, K.G., Boyd, H., Bunskoeke, E.J. & Jukeema, J. 2005. Demography of Eurasian Golden Plovers *Pluvialis apricaria* staging in the Netherlands, 1949–2000. *Ardea* 93: 49–64.

Piiroinen, J., Tiainen, J., Pakkala, T. & Ylimauna, J. 1985. Suomen peltolinnut 1984. *Lintumies* 20: 126–138.

Purroy, F.J. (ed). 1997. *Atlas de las Aves de España 1975–1995*. Lynx Edicions.

Pusey, P. 1843. Evidence on the antiquity, cheapness and efficacy of thorough draining, or land ditching, as practised in the counties of Suffolk, Hertford, Essex and Norfolk. *Journal of the Royal Agricultural Society of England* 4: 23–49.

Pyefinch, R. & Golborn, P. 2001. *Atlas of the Breeding Birds of Lancashire and north Merseyside*. Hobby Publications.

Ratcliffe, D. 1980. *The Peregrine Falcon*. T & AD Poyser. Calton.

Ratcliffe, D. 2005. *Lapland: a natural history*. T & AD Poyser. London.

Redfern, C.P.F. 1982. Lapwing nest sites and chick mobility in relation to habitat. *Bird Study* 29: 201–208.

Redfern, C.P.F. 1983. Aspects of the growth and development of Lapwings *Vanellus vanellus*. *Ibis* 125: 266–272.

Reijnen, R., Foppen, R. & Meeuwsen, H. 1996. The effects of traffic on the density of breeding birds in Dutch agricultural grasslands. *Biological Conservation* 75: 255–260.

Rheinwald, G. 1993. *Atlas der Verbreitung und Haufigkeit der Brutvogel Deutschlands*. Dachverband Deutscher Avifaunisten. Germany.

Rietkerk, F. & Wacher, T. 1996. The birds of Thumamah, Central Province, Saudi Arabia. *Sandgrouse* 18: 24–52.

Rinkel, G.L. 1940. Waarnemingen over het gedrag van de Kievit *Vanellus vanellus* (L) gedurende de broetijd. *Ardea* 29: 108–147.

Riviere, B.B. 1930. *A History of the Birds of Norfolk*. Witherby. London.

Roberts, T.J. 1991. *The Birds of Pakistan*. Oxford University Press.

ROBERTSON, J. & BERG, Å. 1992. Status and population changes of farmland birds in southern Sweden. *Ornis Svecica* 2: 119–130.

ROBINSON, M. 1986. The extent of farm under-drainage in England and Wales prior to 1939. *Agricultural History Review* 34(1): 79–85.

ROBINSON, M. & ARMSTRONG, A.C. 1988. The extent of agricultural field drainage in England and Wales, 1971–80. *Trans. Inst. Br. Geogr.* NS 13: 19–28.

ROBINSON, R.A., WILSON, J.D. & CRICK, H.P.Q. 2001. The importance of arable habitat for farmland birds in grassland landscapes. *Journal of Applied Ecology* 38: 1059–1069.

ROBSON, C. 2000. *A Field Guide to the Birds of SE Asia.* New Holland Publishers (UK). London.

ROBSON, R.W. & WILLIAMSON, K. 1972. The breeding birds of a Westmorland farm. *Bird Study* 19: 202–214.

ROGACHEVA, H. 1992. *The Birds of Central Siberia.* Husum. Druck-u. Verlassgesellschaft.

ROJAS DE AZUAJE, L.M., TAI, S. & MCNEIL, R. 1993. Comparison of rod/cone ratio in three species of shorebirds having different nocturnal foraging strategies. *The Auk* 110: 141–145.

ROOTH, J. 1977. 'Meadow bird' populations in the Netherlands. *Bulletin IWRB* 43–4: 76–77.

RSPB. 1983. *Land Drainage in England and Wales: an interim report.* RSPB. Sandy.

RUTTLEDGE, R.F. 1966. *Ireland's Birds.* Witherby. London.

RYVES, B.H. 1948. *Birdlife in Cornwall.* Collins. London.

SALOMONSEN, F. 1967. *Fuglene på Grønland.* Rhodos. Copenhagen.

SALTER, J.H. 1895–1904. Observations on Birds in mid Wales. *Zoologist* 3(19): 179, 3(20): 24–26, 4(8): 66–71.

SAUNDERS, H. 1899. *Illustrated Manual of British Birds.* Van Voorst. London.

SCHIFFERLI, A., GÉROUDET, P. & WINKLER, R. 1980. *Atlas des Oiseaux nicheurs de Suisse.* Sempach Schweizerische Vogelwarte.

SCHMID, H., LUDER, R., NAEF-DAENZER B., GRAF, R. & ZBINDEN, N. 1998. *Schweizer Brutvogelatlas: verbreitung der brutvogel in der Schweiz und in Furstentum Liechtenstein, 1993–1996.* Schweizerische Vogelwarte. Sempach.

SCOTT, D.A. & CARP, E. 1984. Mid-winter survey of the wetlands of Mesopotamia, Iraq, 1979. *Sandgrouse* 4: 60–76.

SEEBOHM, H. 1885. *A History of British Birds.* R.H. Porter. London.

SELOUS, E. 1933. *Evolution of Habit in Birds.* Constable. London.

SEYMOUR, A.S., HARRIS, S., RALSTON, C. & WHITE, P.C.L. 2003. Factors influencing the nesting success of Lapwings *Vanellus vanellus* and behaviour of Red Fox *Vulpes vulpes* in Lapwing nesting sites. *Bird Study* 50: 39–46.

SHEAIL, J. 1976. Land improvement and reclamation: the experiences of the First World War in England and Wales. *Agricultural History Review* 24: 110–125.

SHEAIL, J. & MOUNTFORD, J.O. 1984. Changes in the perception and impact of agricultural land improvement: the post-war trends in Romney Marsh. *Journal of the Royal Agricultural Society of England.* 145: 43–56.

SHELDON, R.D. 2002. *Factors affecting the distribution, abundance and chick survival of the Lapwing* (*Vanellus vanellus*). Ph.D thesis, Harper Adams University College.

SHELDON, R., BOLTON, M., GILLINGS, S. & WILSON A. 2004. Conservation management of Lapwing *Vanellus vanellus* on lowland arable farmland in the UK. *Ibis* 146 (Suppl. 2): 41–49.

SHIRIHAI, H. 1996. *The Birds of Israel.* Academic Press. London.

SHRUBB, M. 1979. *The Birds of Sussex: their present status.* Phillimore. Chichester.

SHRUBB. M. 1985. Breeding habitats of the Lapwing in Sussex. *Sussex Bird Report* 37(1984): 75–80.

SHRUBB, M. 1988. The influence of crop rotations and field size on a wintering Lapwing *V. vanellus* population in an area of mixed farmland in West Sussex. *Bird Study* 35: 123–131.

SHRUBB, M. 1990. Effects of agricultural change on nesting Lapwings *Vanellus vanellus* in England and Wales. *Bird Study* 37: 115–127.

SHRUBB, M. 1993. *The Kestrel.* Hamlyn. London.

SHRUBB, M. 2003. *Birds, Scythes and Combines: a history of birds and agricultural change.* Cambridge University Press. Cambridge.

SHRUBB, M. 2005. Polygyny in Welsh Lapwings. *Welsh Birds* 4: 270–271.
SHRUBB, M. & LACK, P.C. 1991. The numbers and distribution of Lapwings *V. vanellus* nesting in England and Wales in 1987. *Bird Study* 38: 20–37.
SHRUBB, M, WILLIAMS, I.T. & LOVEGROVE, R.R. 1997. The impact of changes in farming and other land uses on bird populations in Wales. *Welsh Birds* 1(5): 4–26.
SIBLEY, C. & MONROE, B. 1990. *Distribution and Taxonomy of Birds of the World.* Yale University Press.
SIBLEY, D.A. 2000. *The Sibley Guide to Birds.* National Audubon Society. A.A. Knopf. New York.
SIM, I.M.W., GREGORY, R.D., HANCOCK, M.H. & BROWN, A.F. 2005. Recent changes in the abundance of British upland breeding birds. *Bird Study* 52: 261–275.
SIRIWARDENA, G.M., BAILLIE, S.R., BUCKLAND, S.T., FEWSTER, R.M., MARCHANT, J.H. & WILSON, J.D. 1998. Trends in the abundance of farmland birds: a quantitative comparison of smoothed Common Birds Census indices. *Journal of Applied Ecology* 35: 24–43.
SMITH, A.C. 1887. *The Birds of Wiltshire.* R.H. Porter. London.
SMITH, A.E. & CORNWALLIS, R.K. 1955. *The Birds of Lincolnshire.* Lincolnshire Naturalists' Union.
SMITH, K.W. 1983. The status and distribution of waders breeding on lowland wet grassland in England and Wales. *Bird Study* 30: 177–192.
SMITH, T. 1917. Twin nests of Lapwing. *British Birds* 11: 46–47.
SMITH, T. 1930–8. The Birds of Staffordshire. *Transactions of the North Staffordshire Field Club,* vols 64–72.
SMOUT, T.C. 2000. *Nature Contested: Environmental History in Scotland and Northern England since 1600.* Edinburgh University Press. Edinburgh.
SMYTHIES, B.E. 1986. *The Birds of Burma (3rd ed).* Nimrod Press. Liss.
SOIKKELI, M. 1965. On the structure of the bird fauna on some coastal meadows in western Finland. *Ornis Fennica* 42: 101–111.
SOIKKELI, M. & SALO, J. 1979. The bird fauna of abandoned shore pastures. *Ornis Fennica* 56: 124–132.
SOLONEN, T. 1985. Agriculture and birdlife in Finland. A review. *Ornis Fennica* 62: 47–55.
SONOBE, K. & ROBSON, J.W. (eds). 1982. *A Field Guide to the Birds of Japan.* Wild Bird Society of Japan.
SOTHERTON, N.W. & SELF, M.J. 2000. Changes in plant and arthropod biodiversity on lowland farmland: an overview. In: Aebischer, N.J., Evans, A.D., Grice, P.V. & Vickery, J.A. (eds) *Ecology and Conservation of Farmland Birds.* British Ornithologists Union. Tring.
SOUTHWELL, T. 1879. Introduction to Lubbock, R. 1879. *Observations on the Fauna of Norfolk.* 2nd edition edited by T. Southwell. Jarrold & Sons. Norwich.
SOVON. 1987. *Atlas van de Nederlandse vogels.* SOVON. Arnhem.
SOVON. 2000. *Atlas de Nederlandse Broevogels.* SOVON. Arnhem.
SPAGNESI, M. & SERRA, L. (eds). 2003. *Uccelli d'Italia.* Quaderni di Conservazione della Natura No.16.
SPENCER, K.G. 1953. *The Lapwing in Britain.* Brown & Sons. London and Hull.
STEVENSON, H. 1870. *The Birds of Norfolk, vol. II.* Van Voorst. London.
STROUD, D.A., DAVIDSON, N.C., WEST, R., SCOTT, D.A., HAANSTRA, L., THORUP, O., GANTER, B. & DELANEY, S. (compilers). 2004. *Status of migratory wader populations in Africa and Western Eurasia in the 1990s.* International Wader Studies 15. International Wader Study Group.
SUFFOLK BIRDS 1998–2001. Suffolk Naturalists' Society.
SULTANA, J., GAUCI, C. & BEAMAN, M. 1975. *A Guide to the Birds of Malta.* Malta Ornithological Society.
SUMMERS, R.W., UNDERHILL, L.G., PEARSON, D.J. & SCOTT D.A. 1987. *Wader migration systems in southern and eastern Africa and western Asia.* In: Davidson, N.C. & Pienkowski, M.W. (eds). The Conservation of International Flyway Populations of Waders. *Wader Study Group Bulletin* 49, Suppl: 15–34.
SUSSEX BIRD REPORT 2000. Sussex Ornithological Society.
SUTHERLAND, W.J. 2004. A blueprint for the countryside. *Ibis* 146 (Suppl. 2): 230–238.
SUTTON, J.E.G. 1966. Ridge and furrow in Berkshire and Oxfordshire. *Oxoniensa* 29–30: 99–115.

SVENSSON, L., GRANT, P.J., MULLARNEY, K. & ZETTERSTROM, D. 1999. *Collins Bird Guide*. HarperCollins. London.

SWINDELLS, D. 1997. Reductions in the numbers of ground-nesting species in Upper Nidderdale. *Harrogate & District Naturalist's Society Bird Report for 1996*: 46–47.

TAPPER, S. 1992. *Game Heritage: an Ecological Review from Shooting and Game-keeping Records*. Game Conservancy. Fordingbridge.

TAYLOR, I.R. & GRANT, M.C. 2004. Long term trends in the abundance of breeding Lapwing *Vanellus vanellus* in relation to land-use change on upland farmland in southern Scotland. *Bird Study* 51: 133–142.

TAYLOR, M., SEAGO, M.J., ALLARD, P. & DORLING, D. 1999. *The Birds of Norfolk*. Pica Press. Mountfield.

TEBBUT, C.F. 1967. *The Birds of Huntingdonshire*. Privately published.

TEIXEIRA, R.M. 1979. *Atlas van de Nederlandse Broedvogels*. Vereniging tot Behoud van Natuurmonumenten in Nederland.

TEUNISSEN, W., SCHEKKERMAN, H. & WILLEMS, F. 2005. *Predatie bij weidevogels* (in Dutch with English summary). Sovon Vogelonderzoek Nederland.

THÉVENOT, M., VERNON, R. & BERGIER, P. 2003. *The Birds of Morocco*. BOU. Tring.

THIBAULT, J-C. & BONACCORSI, G. 1999. *Birds of Corsica*. BOU Checklist 17. BOU. Tring.

THOM, V.M. 1986. *Birds in Scotland*. T & AD Poyser. Calton.

THOMAS, J.F. 1938. Incubation periods of Lapwing and Redshank. *British Birds* 32: 48–49.

THOMAS, J.F. 1939. Incubation period of some waders and fledging period of Lapwing. *British Birds* 33: 85.

THOMPSON, P.S., BAINES, D., COULSON, J.C. & LONGRIGG, G. 1994. Age at first breeding, philopatry and breeding site-fidelity in the Lapwing *Vanellus vanellus*. *Ibis* 136: 474–484.

THOMSON, A.L. (ed). 1964. *A New Dictionary of Birds*. Nelson. London.

THOMSON, D.L. & COTTON, P.A. 2000. Understanding the decline of the British population of Song Thrushes *Turdus philomelos*. In: Aebischer, N.J., Evans, A.D., Grice, P.V. & Vickery, J.A. (eds). *Ecology and Conservation of Farmland Birds*. British Ornithologists' Union. Tring.

THORUP, O. 1998. (The breeding birds on Tipperne 1928–1992). In Danish with English summary. *Dansk Ornitologisk Forenings Tidsskrift* 92: 1–192.

THORUP, O. (compiler). 2005. Breeding waders in Europe, 2000. *International Wader Studies 14*. International Wader Study Group UK.

TIAINEN, J., PAKKALA, T., PIIROINEN, J., VICKHOLM, M. & VIROLAINEN, E. 1985. (Changes in the avifauna of farmland at Lammi, southern Finland, during the past 50 years). *Lintumies* 20: 30–42. English summary.

TICEHURST, C.B. 1932. *A History of the Birds of Suffolk*. Gurney & Jackson. London.

TICEHURST, N.F. 1909. *A History of the Birds of Kent*. Witherby. London.

TINARELLI, R. & BACETTI, N. 1989. Breeding waders in Italy. *Wader Study Group Bulletin* 56: 7–15.

TOMIALOJC, L. 1987. Breeding waders in Poland—their past and present status. *Wader Study Group Bulletin* 51: 38–41.

TOMKOVICH, P.S. 1992. Breeding range and population changes of waders in the former Soviet Union. *Brit. Birds* 85:344–365.

TOOK, G.E. 1936. Lapwing driving sheep from nest. *British Birds* 30: 90.

TRAFFORD, B.D. 1970. Field drainage. *Journal of the Royal Agricultural Society of England* 131: 129–152.

TROLLIET, B. 2003. Elements for a Lapwing (*Vanellus vanellus*) management plan. *Game and Wildlife Science* 20: 93–144.

TROLLIET, B. 2004. *Recensement de Vanneaux et Pluviers dorés Bretagne—Pays de la Loire Janvier 2004*. Report for the Office National de la Chasse et de la Faune Sauvage.

TROLLIET, B. 2005. *Recensement de Vanneaux et Pluviers dorés Janvier 2005: Bretagne et Pays de la Loire, Basse et Haute Normandie, Centre, Poitou-Charente*. Report for the Office National de la Chasse et de la Faune Sauvage.

TUBBS, C.R. 1986. *The New Forest*. Collins. London.

TUBBS, C.R. 1993. An introduction to Hampshire. In: *Birds of Hampshire*. Hampshire Ornithological Society.

TUCKER, G.M. 1992. Effects of agricultural practices on field use by invertebrate-feeding birds in winter. *Journal of Applied Ecology* 29: 779–790.

TUCKER, G.M. & EVANS, M.I. 1997. *Habitats for birds in Europe: a conservation strategy for the wider environment*. BirdLife Conservation Series No: 6. BirdLife International. Cambridge.

TULLY, H. 1948. Share of male and female Lapwing in incubation. *British Birds* 41: 355–356.

TURNER, E.L. 1924. *Broadland Birds*. Country Life. London.

ULFSTRAND, S. & HÖGSTEDT, G. 1976. (How many birds breed in Sweden). *Anser* 15: 1–32. English summary.

URBAN, E.K., FRY, C.H. & KEITH, S. 1996. *The Birds of Africa vol II*. Academic Press. London.

USSHER, R.J. & WARREN, R. 1900. *The Birds of Ireland*. Gurney & Jackson. London.

VAN EERDEN, M.R., PROP, J. & VEENSTRA, K. 1979. The development of the breeding populations in several species of birds in the Lauwerszee-area in eight years following reclamation. *Limosa* 52: 176–190.

VAN IMPE, J. 1988. Een vergelijkend onderzoek narr de broedbiologie van de Kievit *Vanellus vanellus* op braak terrien en op landbouwterrein. *De Giervalk* 78: 287–314.

VAN DER WINDEN, J., GERRITSEN, G. & VAN ROOMEN, M. 1997. Preliminary results of a country-wide survey of staging Golden Plover, Lapwing and Curlew in the Netherlands. *Wader Study Group Bulletin* 86: 31–32.

VAN DER ZANDE, A.N., TER KEURS, J. & VAN DER WEIJDEN, W.J. 1980. The impact of roads on the densities of four bird species in an open field habitat. *Biological Conservation* 18: 299–321.

VARJO, U. 1984. Changes in farming in Finland in 1969–1973. *Fennia* 162: 103–115.

VAURIE, C. 1965. *The Birds of the Palaearctic Fauna: Non-Passeriformes*. Witherby. London.

VELASCO, T. & ALBERTO, L.J. 1993. Numbers, main localities and distribution maps of waders wintering in Spain. *Wader Study Group Bulletin* 70: 33–41.

VENABLES, L.S.V. & VENABLES, U.M. 1955. *Birds and Mammals of Shetland*. Oliver & Boyd. Edinburgh.

VEPSÄLÄINEN, K. 1968. The effect of the cold spring 1966 upon the Lapwing *Vanellus vanellus* in Finland. *Ornis Fennica* 45: 33–47.

VERE BENSON, S. 1970. *Birds of Lebanon and the Jordan area*. I.C.B.P. Cambridge.

VICKERY, J.A., TALLOWIN, J.T., FEBER, R.E., ASTERAKI, E.A., ATKINSON, P.W., FULLER, R.J. & BROWN, V.K. 2001. The management of lowland neutral grasslands in Britain: effects of agricultural practices on birds and their food resources. *Journal of Applied Ecology* 38: 647–664.

VIKSNES, J. (ed). 1989. *Latvijas Ligzdojoso Putnu Atlants 1980–1984*. Riga. In Latvian, English and Russian.

VILLAGE, A. & WESTWOOD, N.J. 1994. The relationship of lapwing *Vanellus vanellus* numbers and feeding rates to earthworm numbers in arable and pasture fields in autumn and winter. In: Tucker, G.M., Davies, S.M. & Fuller, R.J. (eds). *The ecology and conservation of lapwings Vanellus vanellus*. UK Conservation No. 9. JNCC. Peterborough.

VILALTA, A.M. 1985. Wintering waders in the Ebro delta. *Wader Study Group Bulletin* 43: 25–26.

VISSER, G.H. & BEINTEMA, A.J. 1991. Reproductive characteristics of meadow birds and other European waders. *Wader Study Group Bulletin* 61 Supplement: 6–11.

VON HAARTMAN, L. 1973. Changes in the Breeding Bird fauna of north Europe: in Farner, D.S. (ed). *Breeding biology of birds*. National Academy of Sciences. Washington.

VOOUS, K.H. 1960. *Atlas of European Birds*. Nelson. London.

VOOUS, K.H. 1962. De Kievitensterfte in der droge zumer van 1959 *Ardea* 50: 147–161.

WAKEHAM-DAWSON, A. & SMITH, K.W. 2000. Birds and lowland grassland management practices in the UK: an overview. In: Aebischer, N.J., Evans, A.D., Grice, P.V. & Vickery, J.A. (eds). *Ecology and Conservation of Farmland Birds*. British Ornithologists' Union. Tring.

WALASZ, K. & MIELCZAREK, P. (eds). 1992. *Atlas Ptakow Legowych Malopolski 1985–1991*. Biologica Silesiae Wroclaw. In Polish and English.

WELSH BIRD REPORT 1990–2003. Welsh Ornithological Society.

WERNHAM, C., TOMS, M., MARCHANT, J., CLARK, J., SIRIWARDENA, G. & BAILLIE, S. (eds). 2002. *The Migration Atlas: movements of the birds of Britain and Ireland.* T & AD Poyser. London.

WETLANDS INTERNATIONAL 2002. *Waterbird population estimates.* Wetlands International Global series No.12. Wageningen. Netherlands.

WHITE, G. 1789. *The natural history and antiquities of Selbourne.* B. White & Son. London.

WHITTINGHAM, M.J. & EVANS, K.L. 2004. The effects of habitat structure on predation risk of birds in agricultural landscapes. *Ibis* 146 (Suppl.2): 210–220.

WICKENS, J.C. 1948. Behaviour of Lapwings with eggs and young. *British Birds* 41: 28–29.

WILLIAMS, I. 1999. Lapwing nesting on a school roof. *Welsh Birds* 2: 137.

WILLIAMS, G., HENDERSON, A., GOLDSMITH, L. & SPREADBOROUGH, A. 1983. The effects on birds of land drainage improvements in the North Kent Marshes. *Wildfowl* 34: 33–47.

WILLIAMS, G. & HALL, M. 1987. The loss of coastal grazing marshes in South and East England, with special reference to East Essex, England. *Biological Conservation* 39: 243–253.

WILLIAMSON, K. 1945. Some new and scarce breeding species in the Faeroe Islands. *Ibis* 85: 550–558.

WILLIAMSON, K. 1975. Birds and climatic change. *Bird Study* 22: 143–164.

WILLIAMSON, K. & BOYD, J.M. 1963. *A Mosaic of Islands.* Oliver & Boyd. Edinburgh & London.

WILSON, A.H.R. 1926. Late nesting of Lapwing. *British Birds* 20: 298.

WILSON, J. 1967. Trigamy in Lapwing. *British Birds* 60: 217.

WILSON, J.R. 1978. Agricultural influences on waders nesting on the South Uist machair. *Bird Study* 25: 198–206.

WILSON, A.M. & BROWNE, S.J. 1999. Breeding population estimates for Lapwing, Oystercatcher and Curlew in Scotland: results of the 1998 BTO Lapwing survey. *Scottish Birds* 20:73–80.

WILSON, A.M., VICKERY, J.A. & BROWNE, S.J. 2001. Numbers and distribution of Northern Lapwings *Vanellus vanellus* breeding in England and Wales in 1998. *Bird Study* 48:2–17.

WILSON, A.M., AUSDEN, M. & MILSOM, T.P. 2004. Changes in breeding wader populations on lowland wet grasslands in England and Wales: causes and potential solutions. *Ibis* 146 (Suppl. 2): 32–40.

WILSON, A.M., VICKERY, J.A., BROWN, A., LANGSTON, R.H.W., SMALLSHIRE, D., WOTTON, S. & VANHINSBERGH, D. 2005. Changes in the numbers of breeding waders on lowland wet grasslands in England and Wales between 1982 and 2002. *Bird Study* 52: 55–69.

WOLDHEK, S. 1980. *Bird killing in the Mediterranean.* European Committee for the Prevention of Mass Destruction of Migratory Birds. Netherlands.

YARRELL, W. 1845. *A History of British Birds, 2nd edn.* Van Voorst. London.

YEATMAN, L. 1976. *Atlas des Oiseaux nicheurs de France.* Société Ornithologique de France. Paris.

YEATMAN-BERTHELOT, D. 1991. *Atlas des Oiseaux de France en Hiver.* Société Ornithologique de France.

YOUNG, A. 1813. *A General View of the Agriculture of Oxfordshire.* Board of Agriculture. London.

Index

Page numbers in italics refer to illustrations and tables.